Robotic Platforms for Assistance to People with Disabilities

Robotic Platforms for Assistance to People with Disabilities

Editors

Carlos A. Jara
Juan A. Corrales

MDPI • Basel • Beijing • Wuhan • Barcelona • Belgrade • Manchester • Tokyo • Cluj • Tianjin

Editors
Carlos A. Jara
University of Alicante
Spain

Juan A. Corrales
Universidade de Santiago de
Compostela
Spain

Editorial Office
MDPI
St. Alban-Anlage 66
4052 Basel, Switzerland

This is a reprint of articles from the Special Issue published online in the open access journal *Applied Sciences* (ISSN 2076-3417) (available at: https://www.mdpi.com/journal/applsci/special_issues/Robotic_Platforms_Assistance_Disabilities).

For citation purposes, cite each article independently as indicated on the article page online and as indicated below:

LastName, A.A.; LastName, B.B.; LastName, C.C. Article Title. *Journal Name* **Year**, *Volume Number*, Page Range.

ISBN 978-3-0365-3677-4 (Hbk)
ISBN 978-3-0365-3678-1 (PDF)

Contents

About the Editors

Carlos A. Jara received the title of Industrial Engineer in 2005 by the Miguel Hernández University of Elche and received the title of Doctor in 2010 by the University of Alicante. His PhD thesis received the CESEI Award (best Spanish PhD on research in technological applications in the field of education in the areas of Electrical Engineering, Electronics Technology, Telecommunication Engineering and Computer Engineering) in 2011. He is currently an Assistant Professor and researcher at the University of Alicante, and he is member and founder of the Human Robotics group (www.huro.ua.es). He is focused on assistive and rehabilitation robotics, human–robot interaction and design, and the control of robotic platforms for disabled people. He has participated as researcher in several national projects and has coordinated regional projects on the topic of robotic platforms for disabled people (GV/2018/050). He has 29 publications in international journals, among which 25 publications have an impact within the JCR, and 4 of them in journals that are indexed in prestigious databases.

Juan A. Corrales studied Computer Engineering (2005) and completed a a PhD in robotics (2011) at the UA (Spain) based on two lines of study: human–robot interaction for assembly tasks and the planning of robotic hands for dexterous manipulation. Afterwards, he moved to France (UPMC, Paris) to continue his research on robotic manipulation in the FP7 HANDLE project. In 2014, he moved to Clermont-Ferrand where he worked as an Associate Professor at the 'SIGMA Clermont' graduate school and at the 'Institut Pascal' laboratory. He participated in an industrial project (FUI Aerostrip, 2015–18) about human–robot interaction for stripping airplanes and a H2020 project (Bots2Rec, 2016–19) about the multi-modal control of mobile manipulators for asbestos removal. He coordinated (as the technical manager) two European projects: H2020 SoftManBot (2019–21) and SUDOE Commandia (2018–21), both of which were about controlling mobile manipulators for handling deformable objects. In 2021, he began participating in a the H2020 ACROBA project and joined USC (Spain).

Preface to "Robotic Platforms for Assistance to People with Disabilities"

People with congenital and/or acquired disabilities form a significant number of dependents in the current society. These patients lack enough autonomy to live an independent life. Robotic platforms to help people with disabilities are being developed with the aim of providing both rehabilitation treatment and assistance to improve their quality of life and are mainly applied to people who have mobility problems or some type of functional disability. The impact and capacity of assistance of collaborative robotics in this area has continuously improved the healthcare world in terms of chronic disease prevention, saving time for professionals, and lower public health spending. In this sense, something that is important to emphasize in these robotic assistance environments is human–robot interaction. This topic demands sensitive and intelligent robotics platforms that are equipped with complex sensory systems, high handling functionalities, safe control strategies, and intelligent computer vision algorithms.

This Special Issue of *Applied Sciences*, "Robotic Platforms for Assistance to People with Disabilities", aims to cover recent advances in the field of robotic platforms to assist disabled people in daily or clinical environments. Papers address innovative solutions in this field, including affordable assistive robotics devices, new techniques in the control/computer vision for intelligent and safe human–robot interaction, exoskeletons or exosuits to assist people with mobility problems, and advances in mobile manipulators for assistive tasks.

Carlos A. Jara, Juan A. Corrales
Editors

applied
sciences

MDPI

Editorial

Robotic Platforms for Assistance to People with Disabilities

Carlos A. Jara [1,*] **and Juan A. Corrales** [2]

[1] Human Robotics Group, University of Alicante, 03690 Alicante, Spain

[2] Centro Singular de Investigación en Tecnoloxías Intelixentes (CiTIUS), Universidade de Santiago de Compostela, 15782 Santiago de Compostela, Spain; juanantonio.corrales@usc.es

* Correspondence: carlos.jara@ua.es

1. Introduction

People with congenital and/or acquired disabilities constitute a great number of dependents in today's society. These patients lack enough autonomy to live an independent life. Robotic platforms for helping people with disabilities are being developed with the aim of providing both rehabilitation treatment and assistance in improving their quality of life, mainly for those who have mobility problems or some type of functional disability.

A high demand for services of assisted and rehabilitation robotic platforms is expected as a result of the health status of the world due to the COVID-19 pandemic. Currently, according to the WHO (World Health Organization), existing rehabilitation services have been disrupted in 60–70% of countries due to this pandemic, because of the need to avoid human contact. Therefore, countries must face major challenges to ensure the health and autonomy of their disabled population. Robotic platforms are necessary to ensure assistance and rehabilitation for disabled people in the current global situation.

The capacity of robotic platforms in this area must be continuously improved in order to benefit the healthcare sector in aspects such as chronic disease prevention, assistance, and autonomy. For this reason, research about human–robot interaction in these robotic assistance environments must grow and advance because this topic demands sensitive and intelligent robotic platforms, equipped with complex sensory systems, high handling functionalities, safe control strategies, and intelligent computer vision algorithms. All these technological and scientific developments in different aspects of human–robot interaction can also be extended to other application areas (industry, agriculture, education, etc.) where the assistance of robots is crucial due to physically and/or psychologically demanding tasks.

This Special Issue of *Applied Sciences* called "Robotic Platforms for Assistance to People with Disabilities" has published seven papers which cover recent advances in the field of robotic platforms to assist disabled people in daily or clinical environments. The papers address innovative solutions in this field, including affordable assistive robotics devices, new techniques in control/computer vision for intelligent and safe human–robot interaction, and advances in mobile manipulators for assistive tasks. These papers represent significant contributions to the research field, which will be summarized in the next section.

2. Contributions

Mobile robotic platforms are usually employed to support elderly people in living independently and to assist disabled people in carrying out the activities of daily living independently. In [1], Catalan et al. present a modular mobile robotic platform to assist impaired people based on an upper-limb robotic exoskeleton mounted on a robotized wheelchair. This approach, in comparison with the existing robot assistants, can highlight its modularity and capability to help disabled people with many components: voice-control system, eye-tracking glasses, RGB-D cameras, robotic arm and hand exoskeletons, and a BNCI (brain/neural–computer interaction) system.

In most cases, disabled people must be guided through indoor/outdoor places to their destination, avoiding any obstacles. Specifically, visually impaired people must overcome

Citation: Jara, C.A.; Corrales, J.A. Robotic Platforms for Assistance to People with Disabilities. *Appl. Sci.* **2022**, *12*, 2235. https://doi.org/10.3390/app12042235

Received: 7 February 2022
Accepted: 14 February 2022
Published: 21 February 2022

Publisher's Note: MDPI stays neutral with regard to jurisdictional claims in published maps and institutional affiliations.

many difficulties to walk safely to their destination. For that reason, robotic platforms for visually impaired people must integrate a suitable communication method to assist them. In these cases, dialogue in natural language is the most important communication method to guide them. In [2], a dialogue system for human–robot communication is proposed based on the knowledge graph in order to provide an accurate destination for the navigation system.

Human pose estimation is a current topic of research, especially for the safety of human–robot collaboration and the evaluation of human biomarkers. In this field, evaluation of low-cost markerless human pose estimators has received much attention for their diversity of applications, especially in rehabilitation robotic environments. In [3], Hernández et al. present an evaluation of the angles in the elbow and shoulder joints, estimated by OpenPose and Detectron 2, during four typical upper-limb rehabilitation exercises: elbow side flexion, elbow flexion, shoulder extension, and shoulder abduction. A low-cost setup of two Kinect 2 RGB-D cameras was used to obtain the ground truth of the joint and skeleton estimations during the different exercises.

Direct physical assistance from robotic systems can be useful for the rehabilitation of damaged limbs in accidents, or for the prevention of musculoskeletal problems and fatigue in repetitive tasks. Two main approaches have been developed in the literature: exoskeletons that are tightened to the user's body in order to move their limbs directly, and collaborative robots that handle the tools that have to be used by the user in physically demanding tasks. Two main techniques are used for controlling these systems: by interpreting brain signals so that the user's intentions are estimated, or by including force/torque feedback so that the force applied by the user in a tool is processed and amplified. In [4], Ferrero et al. use an EEG (electroencephalogram) cap for controlling a lower-limb exoskeleton based on two phases (training and tests) applied in walking and standing conditions. In [5], Maithani et al. use an impedance control strategy that calculates the force applied by a user in a tool (e.g., a knife), attached to a collaborative robotic arm for meat-cutting applications. They include an intent prediction module in order to reduce the forces applied by the user by 20%, regarding classical force amplification strategies.

In addition to the previous direct physical assistance solutions, rehabilitation applications can also benefit from sensor-based monitoring systems that have lower costs and thus can be used widely, not only in hospitals but also at home. Gomez-Donoso et al. in [6] propose a new software platform that is able to analyze data coming from a low-cost hand-tracking device and a low-cost surface electromyography (sEMG) sensor, in order to verify that a set of rehabilitation exercises is carried out competently. This software is implemented in a social robot in order to improve the engagement of the patients and to improve feedback about the therapy.

The use of monitoring devices can also be applied to neurological disorders in order to improve the living conditions of people suffering from them. In [7], Vicente-Samper et al. propose mixing data from a personal device (that measures motor activity with an inertial sensor, heart rate, and body temperature) with that from an environmental device (based on a camera for tracking people around the user in order to detect social interactions) into a standard database where machine learning algorithms can extract user models (i.e., concentration level—distracted vs. focused—of the user while performing a task, such as reading a book).

Conflicts of Interest: The authors declare no conflict of interest.

References

1. Catalan, J.M.; Blanco, A.; Bertomeu-Motos, A.; Garcia-Perez, J.V.; Almonacid, M.; Puerto, R.; Garcia-Aracil, N. A Modular Mobile Robotic Platform to Assist People with Different Degrees of Disability. *Appl. Sci.* **2021**, *11*, 7130. [CrossRef]
2. Chen, C.H.; Shiu, M.F.; Chen, S.H. Use Learnable Knowledge Graph in Dialogue System for Visually Impaired Macro Navigation. *Appl. Sci.* **2021**, *11*, 6057. [CrossRef]
3. Hernández, Ó.G.; Morell, V.; Ramon, J.L.; Jara, C.A. Human Pose Detection for Robotic-Assisted and Rehabilitation Environments. *Appl. Sci.* **2021**, *11*, 4183. [CrossRef]

4. Ferrero, L.; Quiles, V.; Ortiz, M.; Iáñez, E.; Azorín, J.M. A BMI Based on Motor Imagery and Attention for Commanding a Lower-Limb Robotic Exoskeleton: A Case Study. *Appl. Sci.* **2021**, *11*, 4106. [CrossRef]
5. Maithani, H.; Corrales Ramon, J.A.; Lequievre, L.; Mezouar, Y.; Alric, M. Exoscarne: Assistive Strategies for an Industrial Meat Cutting System Based on Physical Human-Robot Interaction. *Appl. Sci.* **2021**, *11*, 3907. [CrossRef]
6. Gomez-Donoso, F.; Escalona, F.; Nasri, N.; Cazorla, M. A Hand Motor Skills Rehabilitation for the Injured Implemented on a Social Robot. *Appl. Sci.* **2021**, *11*, 2943. [CrossRef]
7. Vicente-Samper, J.M.; Avila-Navarro, E.; Esteve, V.; Sabater-Navarro, J.M. Intelligent Monitoring Platform to Evaluate the Overall State of People with Neurological Disorders. *Appl. Sci.* **2021**, *11*, 2789. [CrossRef]

Article

A Modular Mobile Robotic Platform to Assist People with Different Degrees of Disability

Jose M. Catalan [1,*], Andrea Blanco [1,*], Arturo Bertomeu-Motos [2,*], Jose V. Garcia-Perez [1], Miguel Almonacid [3], Rafael Puerto [1] and Nicolas Garcia-Aracil [1]

[1] Biomedical Neuroengineering Group, Bioengineerig Institute, Universidad Miguel Hernández, 03202 Elche, Spain; j.garciap@umh.es (J.V.G.-P.); r.puerto@umh.es (R.P.); nicolas.garcia@umh.es (N.G.-A.)
[2] Department of Software and Computing Systems, University of Alicante, 03690 San Vicente del Raspeig, Spain
[3] Robotics and Rehabilitation Laboratory, Columbia University, New York, NY 10027, USA; Miguel.Almonacid@upct.es
* Correspondence: jcatalan@umh.es (J.M.C.); ablanco@umh.es (A.B.); arturo.bertomeu@ua.es (A.B.-M.)

Abstract: Robotics to support elderly people in living independently and to assist disabled people in carrying out the activities of daily living independently have demonstrated good results. Basically, there are two approaches: one of them is based on mobile robot assistants, such as Care-O-bot, PR2, and Tiago, among others; the other one is the use of an external robotic arm or a robotic exoskeleton fixed or mounted on a wheelchair. In this paper, a modular mobile robotic platform to assist moderately and severely impaired people based on an upper limb robotic exoskeleton mounted on a robotized wheel chair is presented. This mobile robotic platform can be customized for each user's needs by exploiting its modularity. Finally, experimental results in a simulated home environment with a living room and a kitchen area, in order to simulate the interaction of the user with different elements of a home, are presented. In this experiment, a subject suffering from multiple sclerosis performed different activities of daily living (ADLs) using the platform in front of a group of clinicians composed of nurses, doctors, and occupational therapists. After that, the subject and the clinicians replied to a usability questionnaire. The results were quite good, but two key factors arose that need to be improved: the complexity and the cumbersome aspect of the platform.

Keywords: assistive robotics; multimodal interfaces; robotic exoskeleton

Citation: Catalan, J.M.; Blanco, A.; Bertomeu-Motos, A.; Garcia-Perez, J.V.; Almonacid, M.; Puerto, R.; Garcia-Aracil, N. A Modular Mobile Robotic Platform to Assist People with Different Degrees of Disability. *Appl. Sci.* **2021**, *11*, 7130. https://doi.org/10.3390/app11157130

Academic Editor: Carlos A. Jara

Received: 21 June 2021
Accepted: 29 July 2021
Published: 2 August 2021

Publisher's Note: MDPI stays neutral with regard to jurisdictional claims in published maps and institutional affiliations.

1. Introduction

There is evidence that early and intensive rehabilitation therapies are associated with better functional gains in patients with acquired brain damage [1]. Rehabilitation robots have shown good results in delivering high-intensity therapies and to maximize patients' recovery [2–4]. However, there are some motor functions that cannot be recovered. In this case, assistive robotics have shown good results in assisting patients with acquired brain damage in performing activities of daily living and/or in supporting elderly people in staying active, socially connected, and living independently. Principally, there are two kinds of assistive robotic devices: one of them is based on mobile robot assistants, such as Care-O-bot, PR2, and Tiago, among others; the other one is based on the use of an external robotic arm or a robotic exoskeleton fixed or mounted on a wheelchair.

On the other hand, there is another approach based on the use of: (i) an external robotic arm fixed or mounted on a wheelchair; or (ii) an exoskeleton robotic device. JACO and iARM are two of the most popular external robotic arms fixed or mounted on wheelchairs. Both robotic arms were designed to be mounted on a user's motorized wheelchair; they have six degrees of freedom and can reach objects at a distance of 90 cm [5]. A study on the practical demands of the potential users of external robotic arms and upper limb exoskeletons for assistance with ADLs can be found in [6]. The study concluded that eating and hairdressing, as well as cleaning, handling food, dressing, and moving nearby items were the ADLs that have received relatively high scores regarding the necessity of external robotic arms. The FRIEND robotic platform is an example of a well-known external robotic arm that assists disabled people in performing ADLs. The FRIEND platform, which belongs to the group of intelligent wheelchair-mounted manipulators, is intended to support disabled people with impairments of the upper limbs in ADLs [7]. On the other hand, dressing, toilet use, transfer, wheelchair control, moving nearby items, and handling food have shown high demand for the necessity of upper limb exoskeletons. Kigachi et al. presented a mechanism and control method of a mobile exoskeleton robot for three-degree-of-freedom upper-limb motion assistance (shoulder vertical and horizontal flexion/extension and elbow flexion/extension motion assistance) [8]. In addition, Meng et al. presented a mobile robotic exoskeleton with six degrees of freedom (DOFs) based on a wheelchair [9].

In this paper, a mobile robotic platform for assisting moderately and severely impaired people in performing daily activities and fully participating in society is presented. The mobile robotic platform was based on an upper limb robotic exoskeleton mounted on a robotized wheel chair. The platform is modular and composed of different hardware components: an unobtrusive and wireless hybrid brain/neural–computer interaction (BNCI) system (electroencephalography (EEG) and electrooculography (EOG)) [10], a physiological signal monitoring system, an electromyography (EMG) system, a rugged, small form-factor, and high-performance computer, a robotized wheelchair, RGB-D cameras, a voice control system, eye-tracking glasses, a small monitor, a robotic arm exoskeleton attached to the wheelchair, and a robotic hand exoskeleton including a mechatronic device to control the pronation/supination of the arm. Moreover, the robotic exoskeleton can be replaced with an external robotic device if needed. The platform has open-source software components as well, such as algorithms to estimate the user's intention based on the hybrid BNCI system, to process the user's physiological reactions, to estimate the indoor location and to navigate, to estimate gaze and to recognize objects, to compute 3D objects and mouth pose, to recognize user activity, and a high-level controller to control the robotic exoskeleton or external robotic device and to control the environment and wheelchair control system. The modularity of the presented mobile robotic platform can be exploited by adapting the multimodal interface to the residual capabilities of the disabled person. In particular, the platform can be mainly adapted to three groups of end users with different residual capabilities:

- Group 1: users with residual motor capabilities to control the arm and/or hand, but who need assistance to carry out activities of daily living in an efficient way. In this group of users, residual EMG signals could be used to control a wearable robot to assist in performing ADLs. In addition, the multimodal interface could be composed of a voice semantic recognition system (for users with non-speech disorders) or a wearable EOG system (for users with speech disorders) to tune some parameters of the high-level controller of the wearable robot and to interact with the user control software, a commercial wearable device for physiological signal monitoring, and RGB depth cameras to sense and understand the environment and context to automatically recognize the abilities necessary for different ADLs;
- Group 2: users without functional control of the arm and/or hand and who are unable to speak (due to a speech disorder or aphasia). In this group, the multimodal interface could be composed of a hybrid BMI system to send commands to the high-level control of the wearable robot, a wearable EOG system to interact with the user control software, a commercial wearable device for physiological signal monitoring, and RGB depth cameras to sense and understand the environment and context to automatically recognize the abilities necessary for different ADLs;
- Group 3: users without functional motor control of the arm and/or hand, with speech disorders, and with limited ability to control the movement of their eyes. In this case, the multimodal interface could be composed of a BMI system to send commands to the high-level control of the wearable robot and to interact with the user control

software, a commercial wearable device for physiological signal monitoring, and RGB depth cameras to sense and understand the environment and context to automatically recognize the abilities necessary for different ADLs.

For users belonging to Groups 1 and 2, a set of application scenarios was identified as possible targets for the AIDE system: drinking tasks, eating tasks, pressing a sensitive dual switch, performing personal hygiene, touching another person, and so on. For users, belonging to Group 3, the identified scenarios were related to communication, the control of home devices, and entertainment.

2. Modular Assistive Robotic Platform

The system is a fully autonomous prototype consisting mainly of a robotized wheelchair with autonomous navigation capabilities, a multimodal interface, and a novel arm exoskeleton attached to the wheelchair (Figure 1).

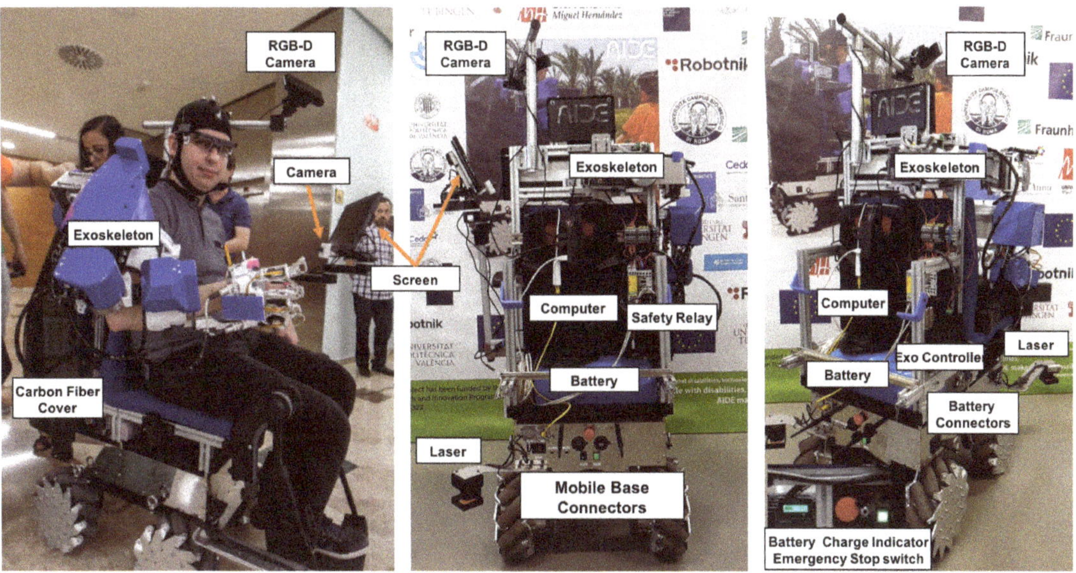

Figure 1. The parts of the platform are shown. The computer, battery, connections with the other components, and the safety relay are located on the back side of the prototype protected by a carbon fiber cover. Safety is a key issue in wearable robotics, so there are three emergency stop switches: (1) on the left side of the robotized wheelchair; (2) on the back side of the robotized wheelchair; and (3) connected through a wire to the left side of the wheelchair.

2.1. Biosignal Acquisition System

The proposed platform is capable of measuring and storing data from several physiological signals. Some of these signals are used for decision making when controlling the system, such as the EOG or EEG, but others are only used to measure the condition of the patient (respiratory rate, galvanic skin response, heart rate, etc.). The system allows adapting the use of the physiological signals based on the patient's need. In addition, new biosignals and processing techniques can be integrated. The performance, signal processing, and adaptation of the different physiological signals of the system have been tested in several studies [11–15].

2.1.1. ExG Cap

An ExG cap, developed by Brain Vision , can be used to perform three different biosignal measurements: (1) EEG acquisition, through eight electrodes, to perform BNCI tasks and allow the user to control the assistive robotic device and interact with the control interface; (2) EOG acquisition, using two electrodes placed on the outer canthus of the eyes, to detect left and right eye movements, to provide the user the opportunity to navigate through the menus of the control interface; (3) EKG acquisition to be combined with the respiration and galvanic skin response (GSR) data in order to estimate the affective state of the user [16].

2.1.2. Electrocardiogram and Respiration Sensor

The system incorporates the Zephyr BioHarness^{TM} (Medtronic Zephyr, Boulder, CO, USA) physiological monitoring telemetry device to measure the electrocardiogram (ECG) and the respiration rate. This device has a built-in signal-processing unit. Therefore, we only applied a 0.004 Hz high-pass filter to remove the DC component of the signals. The HR was extracted from the ECG signal, but the time domain indices of the heart rate variability (HRV) were also extracted. In particular, the SDANN was used as a feature of the HRV, which is defined as the standard deviation of the average instantaneous heart rate intervals (NN) calculated over short periods. In this case, the SDANN was computed over a moving window of 300 s.

2.1.3. Galvanic Skin Response

A GSR sensor, developed by Shimmer, measures the skin conductivity between two reusable electrodes mounted to two fingers of one hand. These data are used, together with the EKG and the respiratory rate, to estimate the affective state of the user [12]. GSR is a common measure in psychophysiological paradigms and therefore often used in affective state detection. The GSR signal was processed using a band-pass filter of 0.05–1.5 Hz (the frequency range of the skin conductance response (SCR)) in order to remove the artifacts.

2.2. Environment Perception and Control System

The system integrates a computer vision system to recognize the environment with which the system will interact [17]. In addition, it has a user interface so that the user can interact with the environment.

2.2.1. Computer Vision System

The activities of daily living (ADLs) require the capability to perform reaching tasks within a complex and unstructured environment. This problem should be solved in real time to be able to deal with the possible disturbances that the element may suffer during the interaction. Moreover, the objects are commonly textureless.

Currently, several methods have been proposed. However, despite the great advances in the field (especially using deep learning techniques), it has not been solved effectively yet, especially with nontextured objects. Some authors have used commercial tracking systems such as Optitrack or ART Track [18–20]. The main limitation of these devices is the necessity to modify the objects to track through the inclusion of optical markers, to reconstruct their position and orientation. The main lines of investigation in the field of 3D textureless object pose estimation are methods based on geometric 3D descriptors, template matching, deep learning techniques, and random forests.

Our system incorporates a computer vision system based on the use of three devices (Figure 1). The first one is Tobii Pro Glasses 2. This eye-tracking system allows the user to select the desired object. The second one is the Orbbec Astra S RGB-D camera used for the 3D pose estimation of textureless objects with which the system can interact. This camera is attached directly to the back of the wheelchair by means of a structure that places it on top of the user's head, focusing on the scene. Finally, a full HD 1080p camera able to work at 30 fps is placed in front of the user, under the screen. This camera is used to estimate the 3D pose of the mouth of the user. This information helps the system know which position the exoskeleton must be in for tasks such as eating or drinking.

This computer vision system was tested in real conditions with patients and was also thoroughly evaluated both qualitatively and quantitatively. The results and a more detailed explanation of the algorithms developed can be seen in [17].

2.2.2. User Interface

The system also has a screen attached to the wheelchair and located in front of the user (Figure 1). On this screen, the interface menus are displayed. It brings many different options to the user (e.g., go to another room, drink, grab an object, entertainment, etc.) and gives some information about the selected task and the exoskeleton status.

2.3. Mobile Platform

The mobile platform was based on Summit XL Steel, from Robotnik. It has omnidirectional wheels that allow the movement of the user in the room. Furthermore, it has its own computer that executes a navigation system, which makes it possible to move between different rooms. Laser-based simultaneous localization and mapping (SLAM) is used to perform the mapping of each room, and the navigation and localization along the different rooms is performed using the adaptive Monte Carlo localization (AMCL) probabilistic localization system, as can be observed in Figure 2). In addition, this platform is equipped with two laser sensors used to provide the wheelchair with an obstacle avoidance algorithm, increasing the safety during navigation.

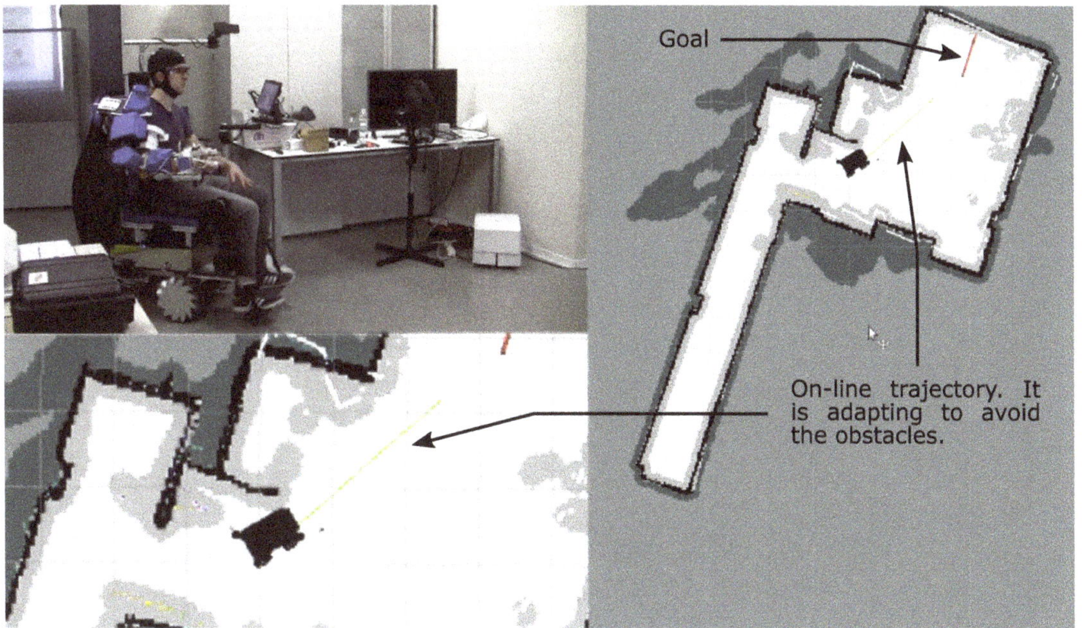

Figure 2. Pictures of the monitoring of the navigation algorithm with obstacle avoidance in a real test.

2.4. Electric Power System

The system has three batteries to power the whole system. First, the mobile platform incorporates a 15/30Ah@48V LiFePO$_4$ battery, which gives an autonomy of up to 10 h. On the other hand, the main computer of the system also has its own 91 kWh battery. The third and the last battery of the system are dedicated to supply the arm and hand exoskeleton. This battery was built with Panasonic 18650b cells and has a capacity of 1.18 kWh, which gives an autonomy of up to 3 h in continuous operation.

2.5. Safety Buttons

Safety is a key issue in wearable robotics, so there are three emergency stop switches (Figure 3): (1) on the left side of the robotized wheelchair; (2) on the back side of the robotized wheelchair; and (3) connected through a wire to the left side of the wheelchair.

By default, there is only one emergency button that kills the exoskeleton power supply from the battery, located on a panel on the wheelchair. However, there is a second plug that offers the possibility of wiring a second button, which allows halting the device from a distance.

To restart the exoskeleton operation after a safety stop, the emergency button must be released and the lit green button of the left panel must be pressed.

The mobile robotic platform has its own emergency button located on the back side of the robotized wheelchair.

To restart the movement of the robotized wheelchair after a safety stop, the platform must be restarted by following the following steps: (1) pressing the green CPU button for 2 s; (2) when the green LED of the CPU button is off, putting the ON-OFF switch in the OFF position; (3) putting the ON-OFF switch in the ON position; this will turn on the platform electronics again; (4) pressing the green CPU button for 2 s; and (5) releasing the safety button.

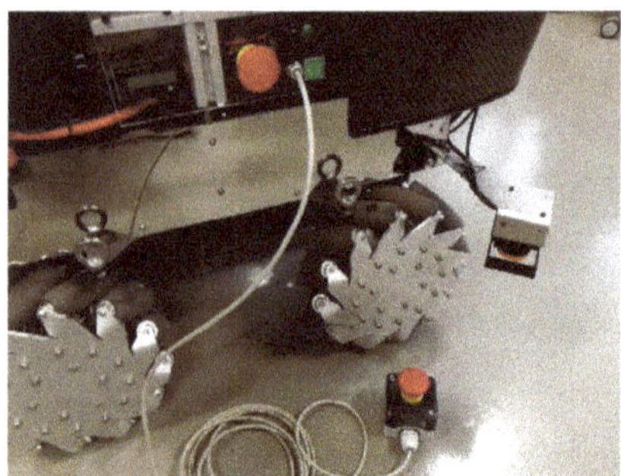 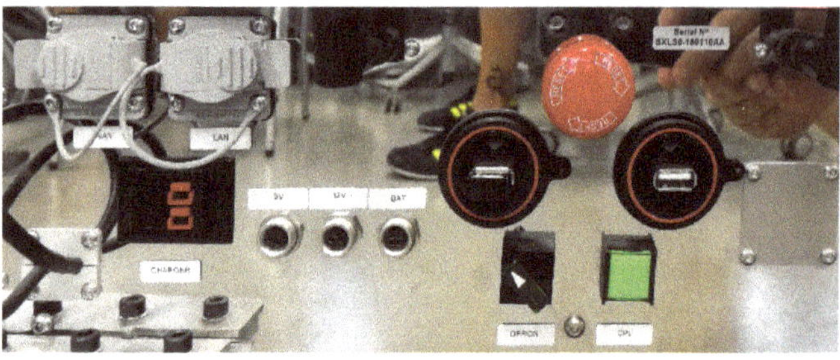

Figure 3. Safety buttons: (**left**) two safety buttons on the left side of the robotized wheelchair for the main CPU; (**right**) one safety button on the back of the wheelchair for the mobile platform.

2.6. Assistive Robotic Devices

The system is able to integrate two different types of robotic devices to assist people with disabilities: (i) an external robotic arm; or (ii) a robotic exoskeleton. Both of them mounted are on the robotized wheelchair.

The control architecture of the robot is independent of the type of robot used as an assistive device. This architecture was implemented in a low layer and a high layer. The low layer implements the low-level control of the robotic device. It implements a joint trajectory controller, which executes the trajectories received by the high-level controller. The other layer corresponds to the high-level controller, which is responsible for managing the communication of the robot with the system, but it also implements a motion planning system. This motion planning system resorts to the learning by demonstration (LbD) method based on the dynamic movement primitives (DMPs) proposed and evaluated in [21].

2.6.1. Exoskeleton Robotic Device

An upper limb exoskeleton was designed with five active degrees of freedom corresponding to the following arm movements: shoulder abduction/adduction, shoulder flexion/extension, shoulder internal/external rotation, elbow flexion/extension, and wrist pronation/supination [11,12,21,22]. This device allows the user's right arm to be moved to reach objects, thus facilitating the performance of ADLs (Figure 1).

In addition to the arm exoskeleton, an active hand exoskeleton was designed to assist the opening and closing of the right/left hand [23,24]. It consists of four independent modules anchored to a hand orthosis that actuate the movements of the thumb, index finger, and middle finger, and jointly move the ring and little finger. The configuration of the hand can be adapted according to the size of the hand.

2.6.2. Robotic Manipulator

The system can also integrate an external robotic manipulator. Experimental tests of the complete system were carried out with JACO robot produced by Kinova (Boisbriand, Canada) [25]. This robotic manipulator is a very light manipulator (4.4 kg for the arm and 727 g for the hand), which can be installed on a motorized wheelchair (right or left) to help people with upper extremity mobility limitations. It has seven degrees of freedom, with a two- or three-finger gripper with a maximum opening of 17.5 cm. The JACO robot is capable of loading objects from 3.5 kg to 4.4 kg, being able to reach objects within a radius of 75 cm.

2.7. Processing and Control System

The system has two computers, the main computer of the system and the computer integrated within the mobile robotic platform (Figures 1 and 4).

The computer of the mobile robotic platform executes the navigation algorithms of the mobile platform using all the information from the sensors. It communicates with the main computer to execute the actions received from the system, as well as to inform the system about the current state during the navigation.

The main computer performs the communication between all the components of the system, processes all the information gathered from the sensors and cameras, and controls the arm and hand exoskeletons. This computer has its own 91 kWh battery.

Both computers communicate through a WiFi router. In this way, we can monitor the operation of the entire system by connecting an external computer to the router.

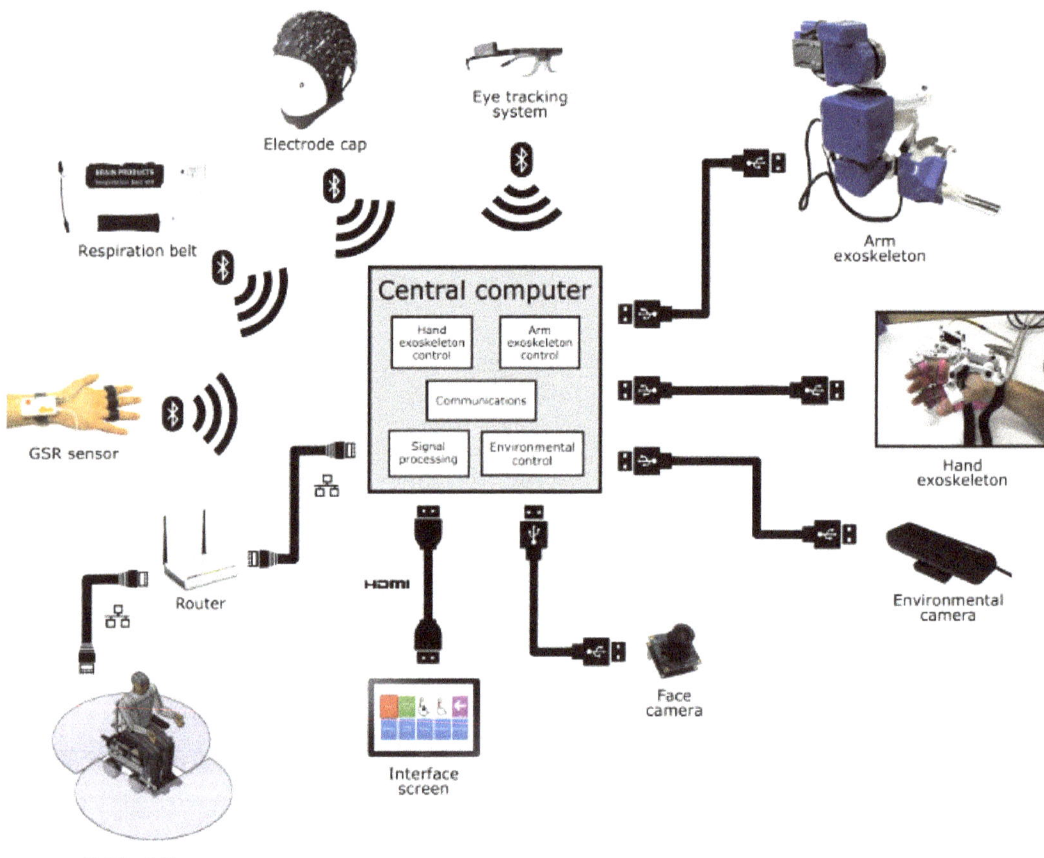

Figure 4. Diagram with the connections between all the components of the platform.

2.8. Finite State Machine

The integration of environmental data acquired by 3D sensors and user intentions has been evaluated in several studies [11–15]. The AIDE system also incorporates an activity recognition algorithm to improve the performance of the control interfaces. This algorithm has been evaluated with patients [16]. The experience gained in these studies resulted in two different state machines (Figures 5 and 6). Both finite state machines (FSMs) describe the general operation of the system, so they have to be adapted according to the user's residual capabilities, in other words depending on the control user interfaces employed. The system can be controlled by means of EEG, EMG, EoG, gaze, voice commands, etc., and/or a combination of these. In this way, the system is adapted to the user's needs or preferences. These FSMs were evaluated in the different studies cited. In addition, in these studies, the different functions of the finite state machines were explained.

2.8.1. Hygiene Task

Due to the complexity of this type of task, the hygiene task is primarily intended to allow the user to be able to clean his/her face or brush his/her teeth. Figure 5 shows the state machine developed to carry out this type of task.

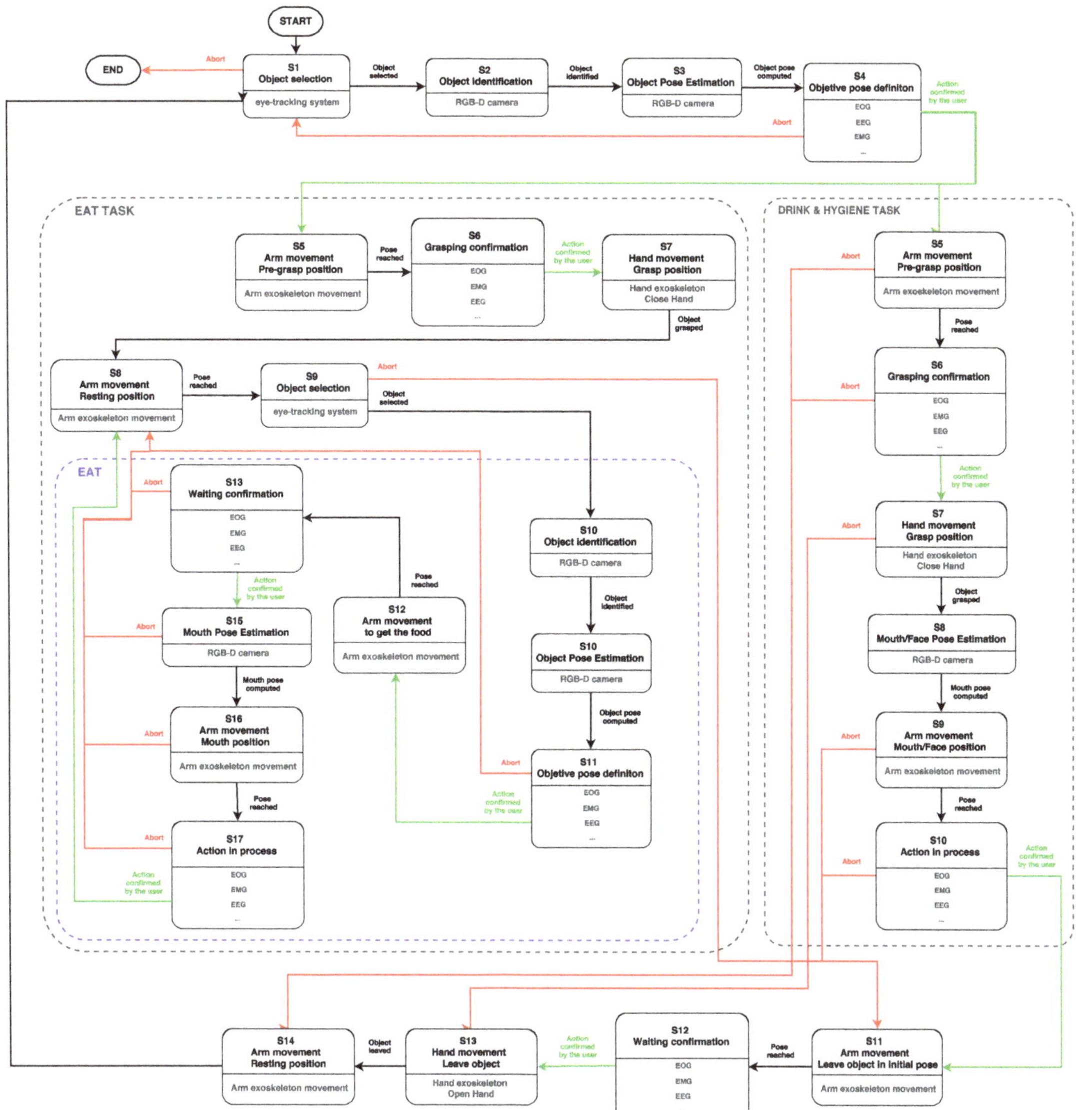

Figure 5. Finite state machine (FSM) for the eating, drinking, and hygiene tasks. They are are sequential implemented allowing the user to continue or abort the task in anytime. Black arrows refer to automatic processes. Green arrow refer to an action confirmed by the user, and the red arrows refer to the decision to abort the current activity by the user.

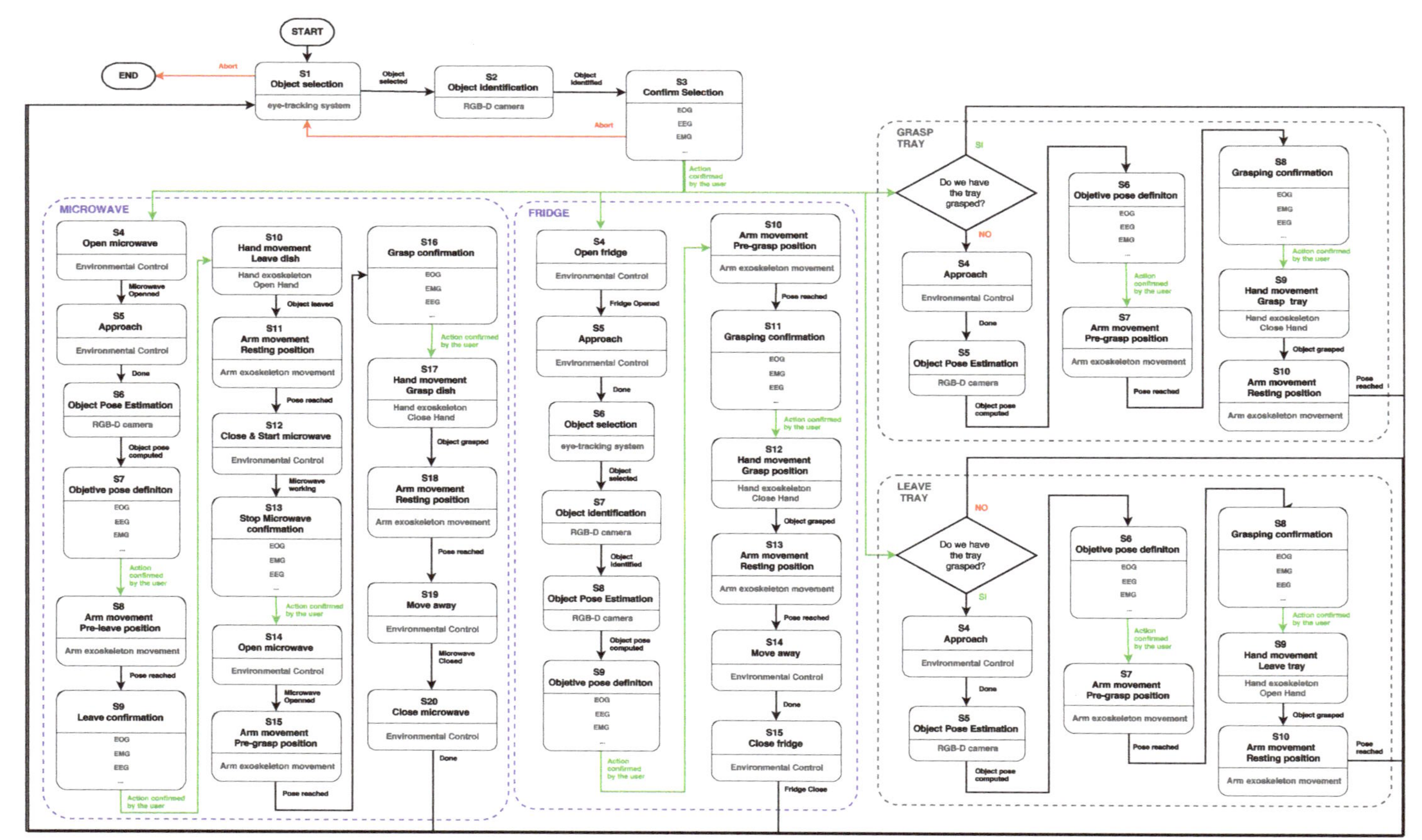

Figure 6. FSM for preparing a meal. The states are sequentially implemented, allowing the user to continue or abort the task at anytime. Black arrows refer to automatic processes. Green arrows refer to an action confirmed by the user, and red arrows refer to the decision to abort the current activity by the user.

2.8.2. Preparing and Eating a Meal

In this scenario, the complex task of preparing and eating a meal is broken up into two subtasks. First, the user has to prepare a meal (Figure 6). In this FSM, the user takes the food from the fridge and heats it in the microwave. To do this, the use moves the wheelchair, opens/closes the fridge, opens/closes the microwave, and moves the robotic arm and hand exoskeleton to grasp and release the food tray. In order to perform this, several elements of the AIDE system are involved such us environmental control to move the wheelchair, the robotic arm, and the hand exoskeleton, the object detection and 3D pose estimation, etc.

After this, the system will continue to the eating and drinking task. In this task, the wheelchair is always in the same position in such a way that the user has only to interact with the exoskeleton to manipulate the glass and the cutlery.

3. Experimental Session

The study presented in this paper aimed to determine the degree of usability of the complete system in its main application environment, assistance in activities of daily living. In other experiments carried out throughout the project [11,12,16,17,21], the different elements that compose the robotic system described here were validated, as well as the different user interfaces used (EEG, EOG, EMG) [13–15].

This experiment was performed in a home environment developed for this purpose. It consisted of a room divided into two areas, one that simulated the living room and the other the kitchen. These two areas were used by a user in order to simulate the interaction with different elements of a home.

For this purpose, we enlisted the collaboration of a subject suffering from multiple sclerosis. In addition, a group of clinicians composed of nurses, doctors, and occupational therapists provided us an objective view of the system in its main field of application after the observation of this experiment (see Figure 7).

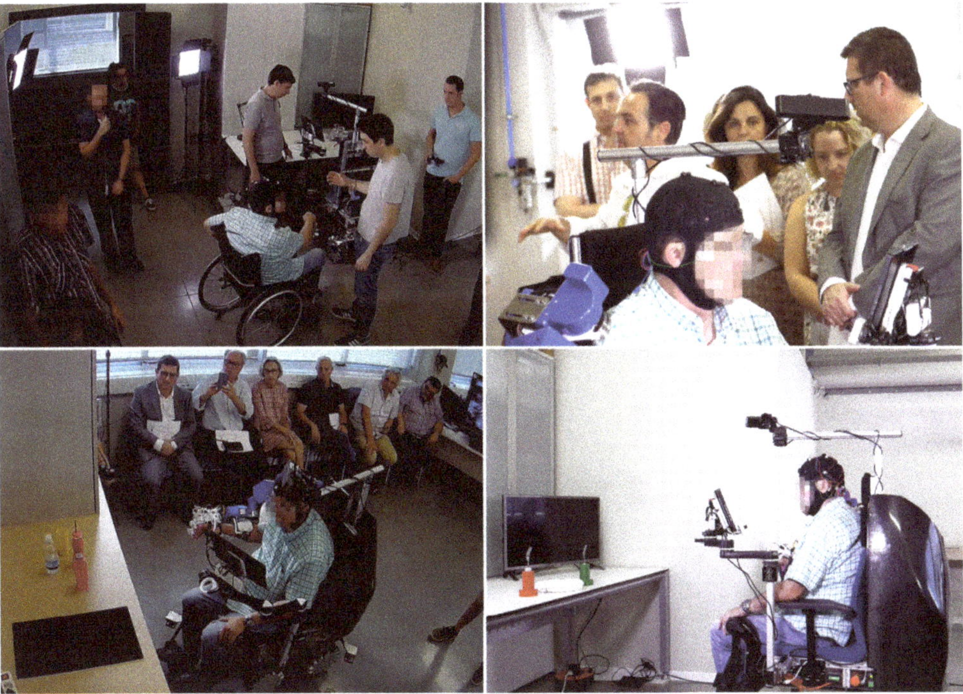

Figure 7. Pictures of the experimental session in the simulated home environment with the subject and the group of clinicians.

The results of this study were obtained by performing the System Usability Scale (SUS), which determines the degree of system usability as perceived by the user and the clinicians.

3.1. Interface

The control of all the system proposed for this experiment was performed through an environmental control interface (ECI). This interface was developed under the AIDE project. It consists of three different abstraction levels where the user has to navigate in order to perform a specific activity (Figure 8). The first level shows the available rooms of the proposed scenario; the second level has a grid with all the possible activities the user can perform; and the last level is related to the action the user can achieve regarding the activity. The control of this interface was performed with a hybrid EEG/EOG system [26]. In addition, the control of the ECI was provided with an intelligent system, proposed in [16], in order to help the navigation through the interface and streamline the completion of the desired task.

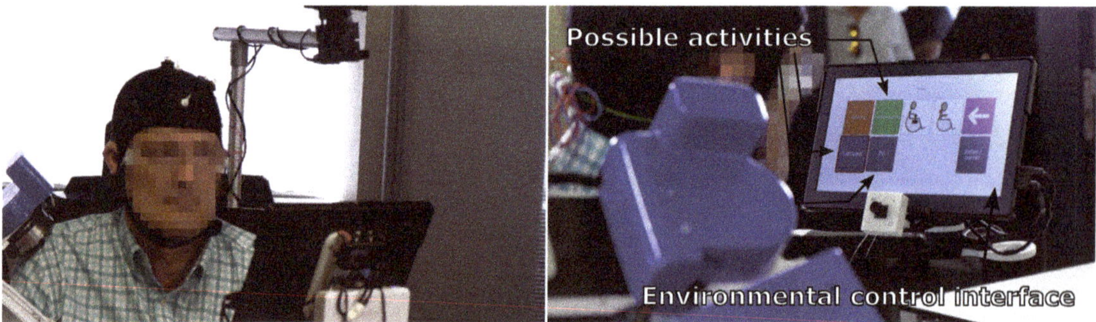

Figure 8. Images of the experimental session where the user navigates through the menus of the control interface to perform the different tasks of the protocol.

3.2. Navigation

In this experiment, two different rooms were mapped, the kitchen and the living room, as can be seen in Figure 9. After a previous mapping of the different rooms, the user could freely navigate through the them using the proposed interface. The navigation to each room was performed in two steps using the interface. First, three different location points were established to perform a direct displacement to them. Then, a fine approach could be performed by small displacements to reach the place where the task had to be executed.

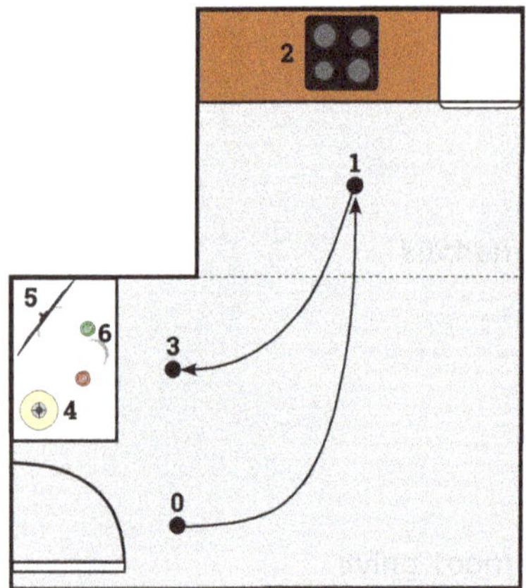

Figure 9. Simulated home scenario.

3.3. Activities of Daily Living

Throughout the experiment, the user interacted with several elements of the home through the use of the environmental control interface (Figure 9). These elements were located in two different rooms, the kitchen and the living room. The user navigated through the environmental control menu using the EOG and EEG interfaces described above.

Environmental control allowed the user to choose the destination he wanted to reach (kitchen or living room), and the mobile platform would take him there automatically. First, as shown in Figure 10, he moved to the kitchen area and adjusted the height of the worktop. Next, he moved to the living room, where he lit a lamp and then turned on the television. The times indicated are those that the user took to complete the activity, from the time he initiated the order to select the task to be performed until the activity was completely finished.

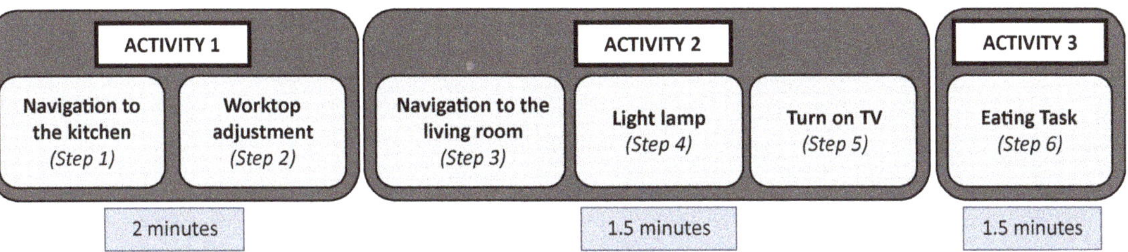

Figure 10. Study protocol.

Once the user had interacted with the different elements of the room, he was ready to perform the eating task. As previously, the user selected the object, in this case the spoon, using the eye-tracking system, and he confirmed the selected object using an EOG command. Therefore, the exoskeleton started to move. When the robot reached the object, the user had to think "close" in order to close the hand (EEG command). When the robot reached his mouth, the user used EOG commands to indicate that he wanted to finish the task or wanted to continue eating. To leave the spoon, the user had to think "open" in order to open the hand (EEG command). At that point, the exoskeleton returned to the idle position, and the finite state machine was left waiting for a new command.

The user was able to complete all the tasks in reasonably short times, since the longest activities were navigation to the kitchen (1 min and 15 s) and the eating task (depending on the repetitions the user wanted to perform). In addition, the user had the ability to abort the activity carried out at any time if he deemed it necessary, providing greater security to the system.

3.4. Subjective Assessment of Usability

The System Usability Scale (SUS) provides a quick tool for measuring the usability aspects of technology. The SUS consists of 10 questions with five response options from strongly agree to strongly disagree. The questions are the following:

Q1 I think that I would like to use this system frequently.
Q2 I found the system unnecessarily complex.
Q3 I thought the system was easy to use.
Q4 I think that I would need the support of a technical person to be able to use this system.
Q5 I found the various functions in this system were well integrated.
Q6 I thought there was too much inconsistency in this system.
Q7 I would imagine that most people would learn to use this system very quickly.
Q8 I found the system very cumbersome to use.
Q9 I felt very confident using the system.
Q10 I needed to learn a lot of things before I could get going with this system.

3.5. Results

As mentioned above, the system developed was validated in different experiments that allowed improving not only the robotic device, but also the control and the different user interfaces. In the study presented in this paper, the main objective was to know the vision of the user himself and the opinion of a group of experts in relation to the usability of the final system in the assistance with ADLs.

To answer the questionnaire, factors such as the time taken by the user to carry out the activity with the robotic system must be taken into account (it cannot be too high), as well as assessing whether the user has completed each of the tasks without problems. To this end, the experts were present as members of the public throughout the experiment, in order to be able to evaluate the aforementioned issues first hand.

All the clinicians filled in the SUS questionnaire, and the results are shown in Figure 11. The median of all the questions was equal to or above 2.5. However, the two questions with the lowest median value were related to the complexity and the cumbersome aspect of the system. This may be due to the fact that this system is a prototype that is still at an early development stage, and it is also a fact that, for the first time of use, it takes a relatively long time to calibrate the control interfaces to the user. We are working on improving the future prototypes of the system by taking into account these aspects.

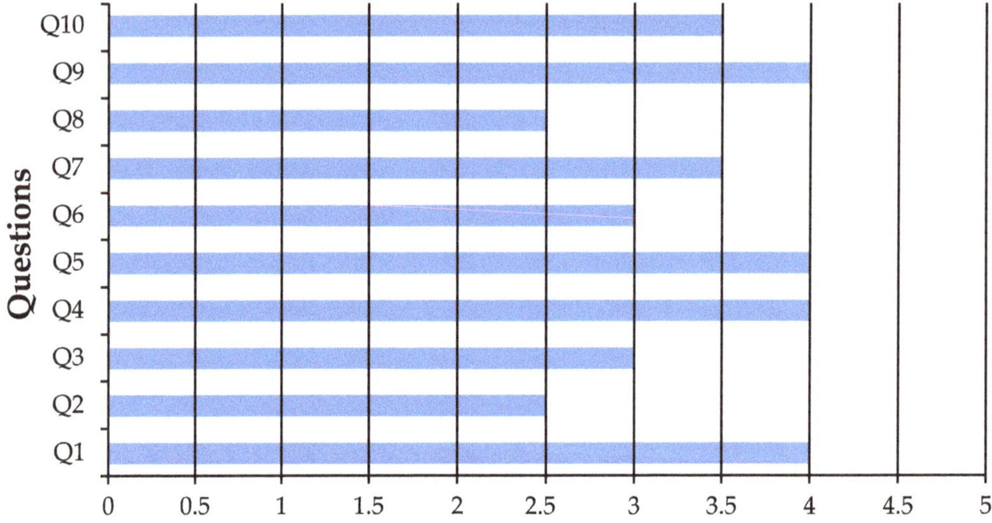

Figure 11. System Usability Scale (SUS) results.

4. Conclusions

In this paper, a modular robotic platform to provide assistance to moderately and severely impaired people in performing daily activities and participating in society was presented. The main innovation of our robotic platform was its modularity, which allows customizing the platform (hardware and software components) for the needs of each potential user. We presented the results of an experiment with a subject suffering from multiple sclerosis. In the experiment, the subject had to carry out different tasks in a simulated scenario while being observed by a a group of clinicians composed of nurses, doctors, and occupational therapists. After that, the subject and the clinicians replied to a usability questionnaire. These results showed a high degree of usability of the system, although there were also several areas for improvement. These aspects were taken into account to improve the new version of the device, thus trying to reduce the users' perception of the complexity of the system.

Author Contributions: J.M.C., A.B., A.B.-M. and J.V.G.-P. designed and developed the platform. J.M.C. and A.B. worked on the construction of the experimental setup. A.B.-M. and J.V.G.-P. performed the experiments. J.M.C. analyzed the data. J.M.C., A.B. and A.B.-M. drafted the paper. J.V.G.-P. deeply revised the manuscript. N.G.-A., M.A. and R.P. contributed to the design of the the the robotic platform and the experiment and deeply revised the manuscript. All the authors checked and approved the final submitted version of the manuscript. All authors read and agreed to the published version of the manuscript.

Funding: This work was supported by the AIDE project through Grant Agreement No. 645322 of the European Commission, by the Conselleria d'Educacio, Cultura i Esport of Generalitat Valenciana, by the European Social Fund—Investing in your future, through the grant ACIF 2018/214, and by the Promoción de empleo joven e implantación de garantía juvenil en I+D+I 2018 through the grant PEJ2018-002670-A.

Institutional Review Board Statement: The study was conducted according to the guidelines of the Declaration of Helsinki and approved by the Ethics Committee of Miguel Hernandez University (2017.32.E.OEP).

Informed Consent Statement: Informed consent was obtained from all subjects involved in the study.

Conflicts of Interest: The authors declare no conflict of interest. The funder institutions had no role in the design of the study; in the collection, analyses, or interpretation of data; in the writing of the manuscript; nor in the decision to publish the results.

Abbreviations

The following abbreviations are used in this manuscript:

AMCL	Adaptive Monte Carlo localization
ADL	Activities of daily living
BNCI	Brain/neural–computer interaction
DMPs	Dynamic movement primitives
ECI	Environmental control interface
EEG	Electroencephalography
EMG	Electromyography
EOG	Electrooculography
FSM	Finite state machine
GSR	Galvanic skin response
LbD	Learning by demonstration
SUS	System Usability Scale
SLAM	Simultaneous localization and mapping

References

1. Turner-Stokes, L.; Pick, A.; Nair, A.; Disler, P.B.; Wade, D.T. Multi-disciplinary rehabilitation for acquired brain injury in adults of working age. *Cochrane Database Syst. Rev.* **2015**, *22*, CD004170. [CrossRef]
2. Chaparro-Rico, B.; Cafolla, D.; Ceccarelli, M.; Castillo-Castaneda, E. Design and Simulation of an Assisting Mechanism for Arm Exercises. In *Advances in Italian Mechanism Science*; Springer International Publishing: Berlin/Heidelberg, Germany, 2017.
3. Rodríguez-León, J.; Chaparro-Rico, B.; Russo, M.; Cafolla, D. An Autotuning Cable-Driven Device for Home Rehabilitation. *J. Healthc. Eng.* **2021**, *2021*, 6680762. [CrossRef] [PubMed]
4. Duret, C.; Grosmaire, A.G.; Krebs, H.I. Robot-Assisted Therapy in Upper Extremity Hemiparesis: Overview of an Evidence-Based Approach. *Front. Neurol.* **2019**, *10*, 412. [CrossRef]
5. Beaudoin, M.; Lettre, J.; Routhier, F.; Archambault, P.; Lemay, M.; Gélinas, I. Impacts of robotic arm use on individuals with upper extremity disabilities: A scoping review. *Can. J. Occup. Ther.* **2018**, *85*, 397–407. [CrossRef]
6. Nam, H.S.; Seo, H.G.; Leigh, J.H.; Kim, Y.J.; Kim, S.; Bang, M.S. External Robotic Arm vs. Upper Limb Exoskeleton: What Do Potential Users Need? *Appl. Sci.* **2019**, *9*, 2471. [CrossRef]
7. Gräser, A.; Kuzmicheva, O.; Ristic-Durrant, D.; Natarajan, S.K.; Fragkopoulos, C. Vision-based control of assistive robot FRIEND: Practical experiences and design conclusions. *at-Automatisierungstechnik* **2012**, *60*, 297–308. [CrossRef]
8. Kiguchi, K.; Rahman, M.H.; Sasaki, M.; Teramoto, K. Development of a 3DOF mobile exoskeleton robot for human upper-limb motion assist. *Robot. Auton. Syst.* **2008**, *56*, 678–691. [CrossRef]

9. Meng, Q.; Xie, Q.; Shao, H.; Cao, W.; Wang, F.; Wang, L.; Yu, H.; Li, S. Pilot Study of a Powered Exoskeleton for Upper Limb Rehabilitation Based on the Wheelchair. *BioMed Res. Int.* **2019**, *2019*, 9627438. [CrossRef] [PubMed]

10. Soekadar, S.R.; Witkowski, M.; Vitiello, N.; Birbaumer, N. An EEG/EOG-based hybrid brain-neural computer interaction (BNCI) system to control an exoskeleton for the paralyzed hand. *Biomed. Eng. Tech.* **2015**, *60*, 199–205. [CrossRef]

11. Crea, S.; Nann, M.; Trigili, E.; Cordella, F.; Baldoni, A.; Badesa, F.J.; Catalán, J.M.; Zollo, L.; Vitiello, N.; Aracil, N.G.; et al. Feasibility and safety of shared EEG/EOG and vision-guided autonomous whole-arm exoskeleton control to perform activities of daily living. *Sci. Rep.* **2018**, *8*, 10823. [CrossRef]

12. Badesa, F.J.; Diez, J.A.; Catalan, J.M.; Trigili, E.; Cordella, F.; Nann, M.; Crea, S.; Soekadar, S.R.; Zollo, L.; Vitiello, N.; et al. Physiological responses during hybrid BNCI control of an upper-limb exoskeleton. *Sensors* **2019**, *19*, 4931. [CrossRef] [PubMed]

13. Nann, M.; Cordella, F.; Trigili, E.; Lauretti, C.; Bravi, M.; Miccinilli, S.; Catalan, J.M.; Badesa, F.J.; Crea, S.; Bressi, F.; et al. Restoring activities of daily living using an EEG/EOG-controlled semiautonomous and mobile whole-arm exoskeleton in chronic stroke. *IEEE Syst. J.* **2020**, *15*, 2314–2321. [CrossRef]

14. Trigili, E.; Grazi, L.; Crea, S.; Accogli, A.; Carpaneto, J.; Micera, S.; Vitiello, N.; Panarese, A. Detection of movement onset using EMG signals for upper-limb exoskeletons in reaching tasks. *J. Neuroeng. Rehabil.* **2019**, *16*, 1–16. [CrossRef]

15. Accogli, A.; Grazi, L.; Crea, S.; Panarese, A.; Carpaneto, J.; Vitiello, N.; Micera, S. EMG-based detection of user's intentions for human-machine shared control of an assistive upper-limb exoskeleton. In *Wearable Robotics: Challenges and Trends*; Springer: Berlin/Heidelberg, Germany, 2017; pp. 181–185.

16. Bertomeu-Motos, A.; Ezquerro, S.; Barios, J.A.; Lledó, L.D.; Domingo, S.; Nann, M.; Martin, S.; Soekadar, S.R.; Garcia-Aracil, N. User activity recognition system to improve the performance of environmental control interfaces: A pilot study with patients. *J. Neuroeng. Rehabil.* **2019**, *16*, 1–9. [CrossRef] [PubMed]

17. Ivorra, E.; Ortega, M.; Catalán, J.M.; Ezquerro, S.; Lledó, L.D.; Garcia-Aracil, N.; Alcañiz, M. Intelligent multimodal framework for human assistive robotics based on computer vision algorithms. *Sensors* **2018**, *18*, 2408. [CrossRef] [PubMed]

18. Onose, G.; Grozea, C.; Anghelescu, A.; Daia, C.; Sinescu, C.; Ciurea, A.; Spircu, T.; Mirea, A.; Andone, I.; Spânu, A.; et al. On the feasibility of using motor imagery EEG-based brain–computer interface in chronic tetraplegics for assistive robotic arm control: A clinical test and long-term post-trial follow-up. *Spinal Cord* **2012**, *50*, 599–608. [CrossRef] [PubMed]

19. Li, M.; Yin, H.; Tahara, K.; Billard, A. Learning object-level impedance control for robust grasping and dexterous manipulation. In Proceedings of the 2014 IEEE International Conference on Robotics and Automation (ICRA), Hong Kong, China, 31 May–7 June 2014; pp. 6784–6791.

20. Ahmadzadeh, S.R.; Kormushev, P.; Caldwell, D.G. Autonomous robotic valve turning: A hierarchical learning approach. In Proceedings of the 2013 IEEE International Conference on Robotics and Automation, Karlsruhe, Germany, 6–10 May 2013; pp. 4629–4634.

21. Lauretti, C.; Cordella, F.; Ciancio, A.L.; Trigili, E.; Catalan, J.M.; Badesa, F.J.; Crea, S.; Pagliara, S.M.; Sterzi, S.; Vitiello, N.; et al. Learning by demonstration for motion planning of upper-limb exoskeletons. *Front. Neurorobot.* **2018**, *12*, 5. [CrossRef] [PubMed]

22. Díez, J.A.; Blanco, A.; Catalán, J.M.; Badesa, F.J.; Sabater, J.M.; Garcia-Aracil, N. Design of a prono-supination mechanism for activities of daily living. In *Converging Clinical and Engineering Research on Neurorehabilitation II*; Springer: Berlin/Heidelberg, Germany, 2017; pp. 531–535.

23. Díez, J.A.; Blanco, A.; Catalán, J.M.; Bertomeu-Motos, A.; Badesa, F.J.; García-Aracil, N. Mechanical design of a novel hand exoskeleton driven by linear actuators. In *Iberian Robotics Conference*; Springer: Berlin/Heidelberg, Germany, 2017; pp. 557–568.

24. Díez, J.A.; Blanco, A.; Catalán, J.M.; Badesa, F.J.; Lledó, L.D.; Garcia-Aracil, N. Hand exoskeleton for rehabilitation therapies with integrated optical force sensor. *Adv. Mech. Eng.* **2018**, *10*. [CrossRef]

25. Campeau-Lecours, A.; Maheu, V.; Lepage, S.; Lamontagne, H.; Latour, S.; Paquet, L.; Hardie, N. Jaco assistive robotic device: Empowering people with disabilities through innovative algorithms. In Proceedings of the Rehabilitation Engineering and Assistive Technology Society of North America (RESNA) Annual Conference, 2016. Available online: https://www.resna.org/sites/default/files/conference/2016/pdf_versions/other/campeau_lecours.pdf (accessed on 1 August 2021).

26. Soekadar, S.; Witkowski, M.; Gómez, C.; Opisso, E.; Medina, J.; Cortese, M.; Cempini, M.; Carrozza, M.; Cohen, L.; Birbaumer, N.; et al. Hybrid EEG/EOG-based brain/neural hand exoskeleton restores fully independent daily living activities after quadriplegia. *Sci. Robot.* **2016**, *1*, eaag3296. [CrossRef]

Article

Use Learnable Knowledge Graph in Dialogue System for Visually Impaired Macro Navigation

Ching-Han Chen [1], Ming-Fang Shiu [1,*] and Shu-Hui Chen [2,3]

[1] Department of Computer Science and Information Engineering, National Central University, Taoyuan City 320317, Taiwan; pierre@g.ncu.edu.tw
[2] Graduate Institute of Learning and Instruction, National Central University, Taoyuan City 320317, Taiwan; anita@ntpu.edu.tw
[3] Center for General Education, National Taipei University, New Taipei City 23741, Taiwan
* Correspondence: 108582003@cc.ncu.edu.tw; Tel.: +886-3-420-7151 (ext. 35211)

Featured Application: Natural language dialogue macro-navigation for the visually impaired. The proposed technology can be applied to other professional fields, such as medical consultation or legal services.

Abstract: Dialogue in natural language is the most important communication method for the visually impaired. Therefore, the dialogue system is the main subsystem in the visually impaired navigation system. The purpose of the dialogue system is to understand the user's intention, gradually establish context through multiple conversations, and finally provide an accurate destination for the navigation system. We use the knowledge graph as the basis of reasoning in the dialogue system, and then update the knowledge graph so that the system gradually conforms to the user's background. Based on the experience of using the knowledge graph in the navigation system of the visually impaired, we expect that the same framework can be applied to more fields in order to improve the practicality of natural language dialogue in human–computer interaction.

Keywords: visually impaired assistance; navigation system; knowledge graph; dialogue system; NLP; reasoning

Citation: Chen, C.-H.; Shiu, M.-F.; Chen, S.-H. Use Learnable Knowledge Graph in Dialogue System for Visually Impaired Macro Navigation. *Appl. Sci.* **2021**, *11*, 6057. https://doi.org/10.3390/app11136057

Academic Editors: Carlos A. Jara and Manuel Armada

Received: 28 February 2021
Accepted: 24 June 2021
Published: 29 June 2021

Publisher's Note: MDPI stays neutral with regard to jurisdictional claims in published maps and institutional affiliations.

1. Introduction

When visually impaired people want to walk safely to their destination, they always have to overcome many difficulties on the street. Nowadays, the common walking aids are still guide dogs and a long cane [1]. Additionally, based on advancements in AI technology, smaller embedded sensors enable wearable devices to effectively detect more road conditions in the surrounding environment [2]. However, in addition to detecting the surrounding environment while walking, visually impaired people also need a macro-navigation that can handle a wide range of information to help plan their travel, such as ticket booking or path planning [3]. Due to the development of Global Positioning Systems (GPS) and Geographic Information Systems (GIS), these technologies are of great help to the development of Electronic Travel Assistance systems (ETA) such as the MOBIC Travel Aid [4], Arkenstone system [5], and Personal Guidance System [6]. However, the use of human–computer interaction to accurately understand the requirements of the user is still in need of substantial improvement [7].

For the visually impaired, the voice is the best way to communicate with a system [8], that is, through Voice User Interfaces (VUI). The latest research on VUI is the Conversational User Interface or Dialogue System, which distinguishes it from other VUI by simulating natural language dialogue instead of command interaction or response interaction [9,10]. Dialogue systems have become the main way of interacting with virtual personal assistants, smart devices, wearable devices, or social robots [11]. Additionally, deep learning technology has also made great contributions to dialogue systems.

Dialogue systems are usually divided into two types, task-oriented and non-task-oriented systems [12]. What we are concerned about here is a task-oriented and multi-turn dialogue system which is suitable for road navigation. Using multi-turn dialogue to understand meaning is the main challenge in this kind of dialogue system. This work focuses on conversation as a means to model context [13] and fully understand the user's intentions.

Understanding the background and making the right response is the main goal of the dialogue system. After parsing the input sentence [14], we recommend using the knowledge graph (KG) as the knowledge base for reasoning dialogue. KG is a way of organizing knowledge. In addition to storing information, it can use deductive methods or inductive methods for reasoning [15,16]. The reasoning process is the way the dialogue system understands the context, and the result of such reasoning becomes the system's response. After each conversation, the system is constantly updated to learn more about the user and provide more accurate results for future applications.

Finally, based on the learnable knowledge graph in the multi-turn dialogue system, and the integration of the widely used GPS and GIS [17], we developed macroscopic walking navigation that can be used by the visually impaired. It can be integrated with micro-navigation to help the visually impaired arrive at targeted goals safely.

2. Methods

Our task-oriented dialogue system is built with a modular architecture. Each module is responsible for a specific task and passes the results to the next module. The modules are Automatic Speech Recognition (ASR), Natural Language Understanding (NLU), Dialogue State Tracking (DST), Dialogue Policy Learning (DPL), Natural Language Generation (NLG), and Text to Speech (TTS). DST and DPL are also called dialogue management. The modular architecture is shown in Figure 1.

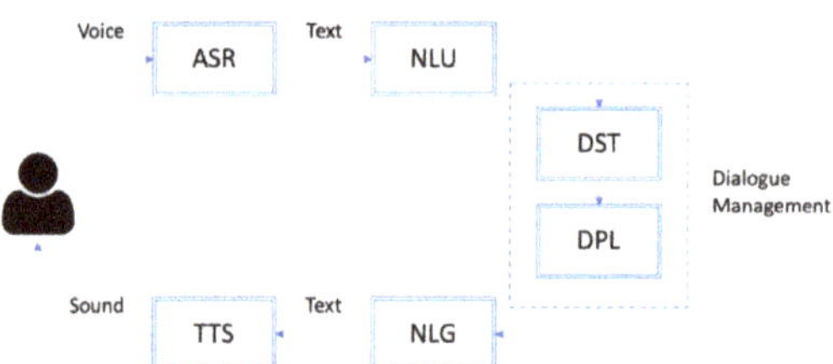

Figure 1. Modular architecture of task-oriented dialogue system.

For ASR and TTS, we use the services provided by the Google Cloud Platform [18]. The functions of the other four main modules are briefly described as follows:

(1) NLU: Maps natural language sentences input by users into machine-readable structured semantic representations.
(2) DST: Tracks users' needs and determines the current conversation status. It integrates the user's current input and all previous conversations in order to understand the meaning by reasoning with context. For dialogue systems, this module is the most significant.
(3) DPL: Determines the action of the system based on the current dialogue state. Also known as Strategy Optimization. The action of the system must conform to the user's intention.
(4) NLG: It transforms the decision of DLP into a natural language to respond to the user, and the voice is sent out by the TTS.

2.1. Knowledge Graph Integration

We propose to use knowledge memory, concept conversion, and logical reasoning of the knowledge graph to do the inference work for DST, and to send the reasoning results to DPL. Our KG uses the RDF triple maps to store information. The triple map is the subject–predicate–object ternary structure [19] and is currently the most mainstream way of storing knowledge graphs. In the understanding of NLU semantics, X-Bar Theory's syntactic analysis theory [20] is used, and the analysis results are converted into triple maps, which are used as the input of the knowledge graph. Figure 2 shows the architecture of the dialogue system after integrating the knowledge graph.

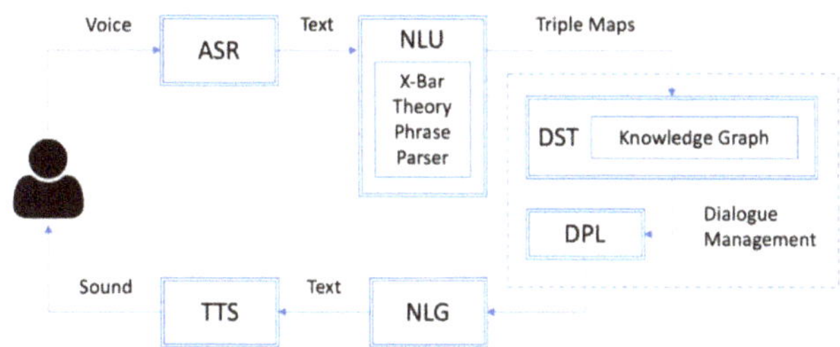

Figure 2. Dialogue system module architecture integrating knowledge graph.

The implementation process of the dialogue system which integrates the knowledge graph is shown in Figure 3. When the user speaks their requirements, the voice is recognized as text through ASR and then passed to the sentence parser of NLU. The sentence parser uses an X-Bar based parsing tool to convert sentences into RDF triple maps, which is the acceptable format for our knowledge graph. RDF triples will be checked for confirmation semantics before being sent to DST for reasoning. If it is a confirmation semantics and the response is affirmative, it means that the user accepts the previous suggestion and agrees to go to the location. Otherwise, the suggestion is canceled. Some details may be ignored (e.g., if no last suggestion was found).

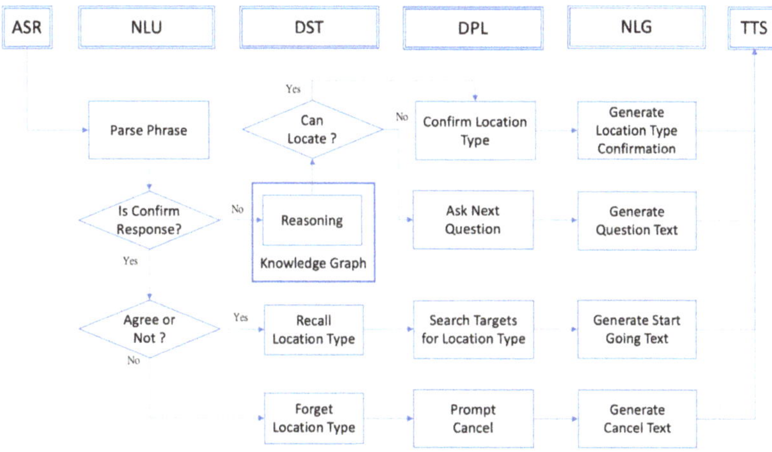

Figure 3. Implementation flowchart of dialogue system integrating knowledge graph.

Unconfirmed semantic triples will enter the knowledge graph of DST for reasoning. The result of the reasoning will take the intention of expression and then decide whether this intention can confirm a specific type of places such as a restaurant or a station. If so, DPL will confirm with the user: "*Do you want to go to the restaurant?*". Otherwise, DPL asks the user how to deal with the intention. For example, when the user says that he is going to have dinner, but dinner is not a type of place, therefore the system will ask the user whether they want to go to a restaurant for dinner.

2.2. Syntax Analysis

We use the Analyzing Syntax of Google Cloud Natural Language API (Google, Mountain View, CA, USA) [21] to analyze the syntax, and convert them into RDF triples maps after obtaining dependency trees based on the X-Bar Theory.

Figure 4 illustrates how to transform dependency trees into RDF triples maps. Taking *"I want to go to Zhishan MRT station"* as an example, we will take the last X-Bars in this phrase, which is (go to) and (Zhishan MRT station), used as the predicate and object, respectively. The *"I"* is as the subject to create a (Person, Traffic, Zhishan MRT station) triplet, and then this triplet will be passed into the knowledge graph for further reasoning processes.

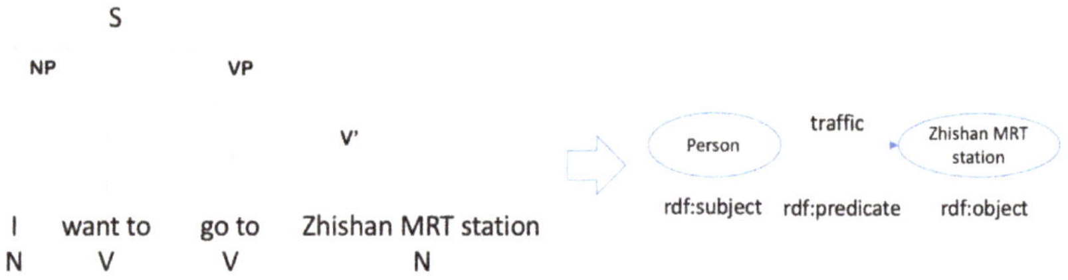

Figure 4. Dependency trees to RDF triple maps.

The process of transforming dependency trees into RDF triple maps includes more details. For example, to adapt to the knowledge graph, the subject *"I"* will be transformed into the upper abstract subject *"Person"*; the verb *"go to"* is also transformed to synonymous predicate *"Traffic"*; and, because the navigation system tends to locate a specific location, but the recipient Zhishan MRT station cannot find lower-level objects in the knowledge graph, it will be directly delivered to DPL to search for it. This concept of conversion via knowledge graph is shown in Figure 5.

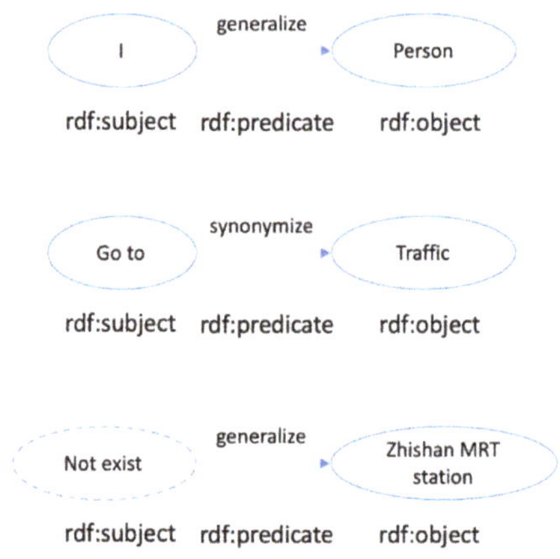

Figure 5. Concept conversion via knowledge graph.

2.3. Reasoning with Knowledge Graph

The reasoning of the knowledge graph mainly revolves around the reasoning of the relationship. Based on the facts or relationships in the graph, it infers unknown facts or relationships [22], and generally pays attention to checking the three aspects of the entity, the relationship, and the structure of the graph. Knowledge graph reasoning techniques are mainly divided into two categories, those based on the deduction (such as description logic [23], Datalog, production rules, etc.) and those based on induction (such as path reasoning [24], representation learning [25], rule learning [26] and reinforcement learning [27], etc.).

This article uses induction-based path reasoning, mainly through the analysis and extraction of existing information in the knowledge graph, since most of the information in the graph represents a certain relationship between two entities. After syntactic analysis, the user's speech is also converted into triples as input so that the two can use triple maps as a communication interface.

We use the PRA (Path Ranking Algorithm) to find the most suitable destination for the user [28], learn the relationship characteristics of the knowledge graph through random walks, quantitatively calculate whether there is a relationship between two nodes, and determine the probability of the relation. The following examples illustrate the application of the PRA algorithm in macro-navigation.

In this case, a visually impaired person wants to go to a restaurant for dinner, but he doesn't know which one to go to, so he says to the navigator, *"I want to have dinner."* The content of dinner in the knowledge graph is shown in Figure 6.

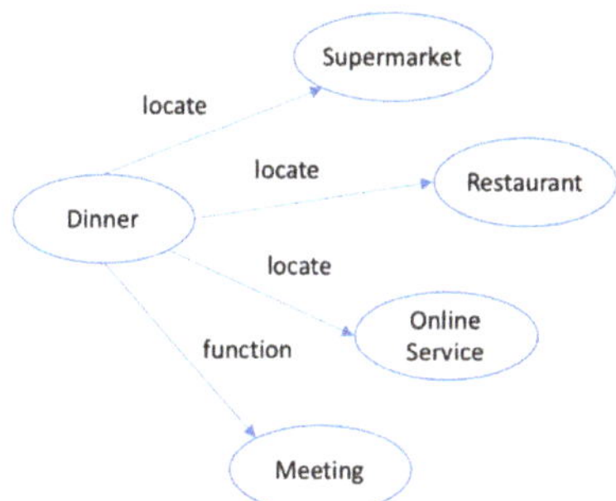

Figure 6. The part of knowledge graph in the case.

Step 1: E_q = {Restaurant, Supermarket, Online Service}, R_1 = locate, for any e $\in$ E_q/R_1, assuming the scoring function h = 1/3, then the following path is shown in Figure 7.

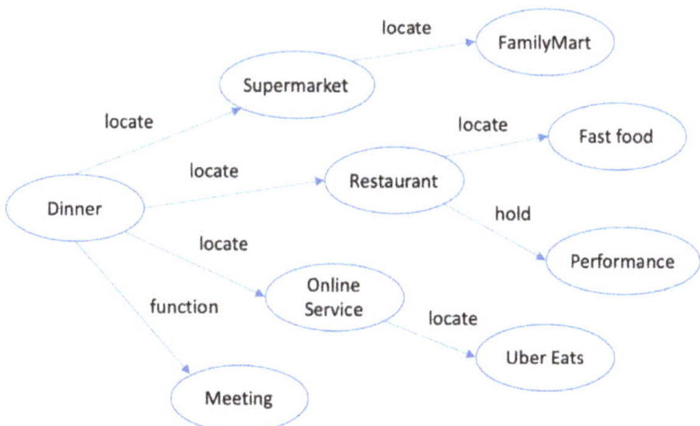

Figure 7. The part of knowledge graph in step 1.

Step 2: E_q = {Restaurant, Fast food}, R_2 = locate, calculate h(Restaurant, locate, Fast food) and h(Restaurant, hold, Performance), obviously h(Restaurant, hold, Performance) = 0. For P_1: Dinner-Restaurant-Fast food, P_2: Dinner-Restaurant-Performance, $h(P_1) > h(P_2)$.
Step 3: And so on, the result is shown in Figure 8.

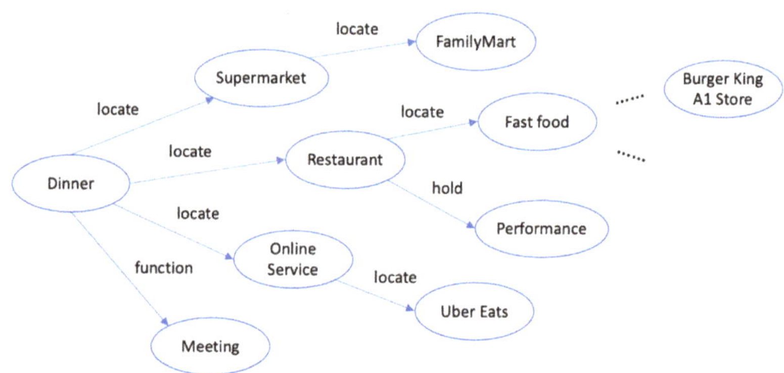

Figure 8. The part of knowledge graph in step 2 and step 3.

Suppose there is a path P: Dinner-Restaurant-Fast food- . . . -Burger King A1 Store, the path length is n, the h_i between two nodes is calculated, and then all h is added to get the entire path P. The score value is $h(P)$.

But it should be noted that the weight of each path is not necessarily the same. For example, the user may prefer to eat McDonald instead of Burger King, so the final score $h(P)$ is given the weight parameter θ, which is also a learnable parameter.

Step 4: Calculate weighted summation.

$$Score(\text{Dinner-}\ldots\text{ - Burger King A1 Store}) = \theta_1 P(1) + \theta_2 P(2) + \ldots + \theta_n P(n) \tag{1}$$

More generally, given a set of paths $P_1, P_2, \ldots, P_n$, one could treat these paths as features for a linear model and rank answers e to the query E_q by

$$\theta_1\, hE_q, P_1\, (e) + \theta_2\, hE_q, P_2\, (e) + \ldots \theta_n\, hE_q, P_n\, (e) \tag{2}$$

The final scoring function:

$$s(e; \theta) = \sum_{P \in \mathcal{P}(q,l)} h_{E_q,P(e)} \theta_P \tag{3}$$

We can construct the training set with the set of relation R and the starting point s and ending point t, and obtain the weight parameter θ through logistic regression. After each conversation, according to the user's decision, the weight parameter will be updated, making the knowledge graph more and more suitable for the user's habits.

3. Results

The main contribution of this paper is to introduce the knowledge graph to the navigation dialogue system and apply the PRA path search algorithm to find the best method for use. We also propose a practical macro-navigation architecture, as shown in Figure 9. The architecture clearly defines the interdependence of the main modules in the dialogue system. In addition to the use of Google Cloud Platform for ASR and TTS, as described above, syntactic analysis is also integrated Google Cloud Natural Language API, and DPL uses Google Maps API to complete the function of geographic path search [29]. In the implementation of the knowledge graph, we use the Apache Jena triplet database (Apache Software Foundation, Forest Hill, MD, USA) [30].

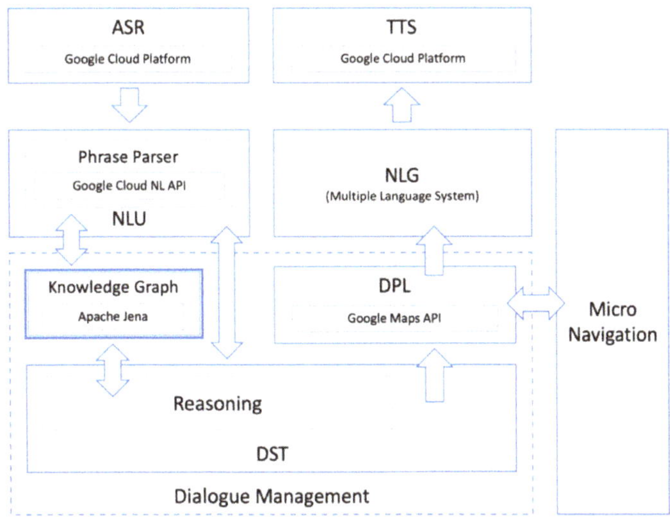

Figure 9. Macro-navigation system architecture based on knowledge graph.

After the user's voice is converted into text by ASR, it is given to Phrase Parser for syntactic analysis. In the process, the support of the knowledge graph will be used to convert phrases into triples in order to provide subsequent path reasoning. Decisions obtained by DST using the results of PRA path reasoning will be executed by DPL, such as using Google Maps API to search for real locations or notify micro-navigation to initiate navigation. The response message processed by DPL will be converted into an appropriate sentence according to the user's language and, finally, sent to TTS to utter a voice to complete a round of dialogue processing.

3.1. Dialogue Experiment

Our experiment is mainly to verify whether the system can discuss an appropriate destination with the user. We designed three scenarios from simple to complex as experimental methods. We use manual methods to create data for the knowledge graph.

The first scenario is that the user directly speaks to a specific destination. The example here is the Seven-Eleven Convenience Store, Xue Cheng branch. This scenario will confirm that the system has the basic ability to command dialogue. The dialogue process is shown in Table 1. Because of the leaf node of the location relationship in the knowledge map of the Seven-Eleven convenience store, Xue Cheng branch, we can see that the system will directly lead the user there.

The system needs a hotword: *"Hi, partner."* to start the dialogue, which makes the dialogue system not too sensitive.

Table 1. Display of the dialogue process of the navigation system in the first scenario.

Speaker	Words	Action
User	Hi, partner.	System activated
Navigator	Hi, is there anything I can help?	
User	I would like to go to Seven-Eleven Xue Cheng Store.	Reasoning
Navigator	Let's go now.	Start micro-navigation

The second scenario is that the user speaks out an indirect destination so that the system can get to the real destination by reasoning. The example here is that the user says that he wants to go to work, and the system deduces that the place where he usually goes to work is his company's location. This scenario will show the ability for simple reasoning, and the dialogue process is shown in Table 2. Since E_q = {Office, Engineering Building 5}, R_1 = locate which has the highest score, the system will advise the user to go to Engineering Building 5.

Table 2. Display of the dialogue process of the navigation system in the second scenario.

Speaker	Words	Action
User	Hi, partner.	System activated
Navigator	Hi, is there anything I can help?	
User	I am going to work.	
Navigator	Your most frequent working place is Engineering Building 5. Are you going there?	Reasoning
User	Okay, please take me there.	
Navigator	Let's go now.	Start micro-navigation

The third scenario is to verify more remote reasoning so that the system can start from a vaguer intention, and obtain the most suitable destination through multiple rounds of dialogue. The scenario here is the same as the description in the previous section. The dialogue process is shown in Table 3.

Table 3. Display of the dialogue process of the navigation system in the third scenario.

Speaker	Words	Action
User	Hi, partner.	System activated
Navigator	Hi, is there anything I can help?	
User	I'm hungry, can you take me to dinner?	
Navigator	Would you like to find a restaurant for dinner?	Reasoning
User	Yes.	
Navigator	Okay, I will look for nearby restaurants for you now.	Search maps
Navigator	There is a Burger King A1 store nearby that you visited last week. Do you want to go?	Found target
User	No, can I go to a Chinese restaurant?	
Navigator	Okay, I will look for a nearby Chinese restaurant for you now.	Search maps
Navigator	There is a 4-star Manfulou restaurant nearby. Are you going to go?	Found target
User	Okay, please take me there.	
Navigator	Let's go now.	Start micro-navigation

3.2. Outdoor Test

We chose a starting point about 300 m away from the Taoyuan A17 MRT station and asked an average blindfolded person and a visually impaired person to conduct the test. The tester walked with the navigation device, communicated with the device, and followed the instructions moving forward. The travel route and dialogue content of the entire use process are shown in Table 4.

First, the tester stated he wanted to take a trip, and after communicating with the navigation system, he decided to go to A17 MRT. Still, shortly after departure, the tester said he wanted to go to the store to buy a drink. Thus, the navigation system took him to the nearest convenience store. Soon, the tester would have liked to leave the convenience store and go to the MRT again. He informed the navigator to take out the previous location and take him there.

Table 4. The walking route and conversation of the entire navigation.

Speaker	Words	Location
User	Hi, partner.	
Navigator	Is there anything I can help?	
User	Please take me to the MRT station.	
Navigator	Okay, let me check, please wait.	
Navigator	There is Dayuan Pilot Station 200 m nearby. May I take you there?	
User	Yes, that's right.	
Navigator	(DPL drives the micro-navigation) Please walk straight along the sidewalk.	
User	Hi, partner.	
Navigator	Hello, how can I help you?	
User	I want to go to the convenience store first.	
Navigator	Okay, let me check, please wait.	
Navigator	There is a FamilyMart convenience store 50 m ahead. You are on your way to the MRT station. Do you want to go to that convenience store first?	
User	Okay, please take me there.	
Navigator	You have arrived at FamilyMart convenience store, please let me know if you want to continue to the MRT station. (Later)	
User	Hi, partner.	
Navigator	Hello, how can I help you?	
User	I am going to the MRT station.	
Navigator	Okay, now I will continue to take you to Dayuan Pilot Station.	

The experiences of the two testers are described as follows.

The test of ordinary people with blindfolded eyes is very successful, even if their eyes are completely covered. Ordinary people do well in communicating with machines and have full confidence in walking. He can reach his destination with little help. It may be that they are already familiar with the system using vision.

Visually impaired people will be extra careful when using equipment in walking. Their step distance will be smaller than that of ordinary people. In conversation with the machine, they will often encounter situations where they cannot talk. The main reason is that they cannot understand the speaking of navigation and must do more training to use the system smoothly. The whole process requires more assistance to reach the destination.

4. Discussion

Our dialogue system can work well in outdoor navigation for visually impaired people. The knowledge graph provides the main contribution. The required navigation information can be correctly understood based on the context and assisted in language generation. This dialogue system provides useful help for visually impaired people walking outdoors.

In the real-world test, we also encountered traditional problems. In a noisy outdoor environment, ASR is very susceptible to environmental noise and cannot accurately obtain the speaking content by the tester, resulting in syntax analysis errors. Besides using noise reduction technology to improve the performance of ASR, we should also enhance the exceptions handling or develop a text correction system based on the knowledge graph.

Our macro-navigation uses many Google cloud services, including ASR, TTS, NLU, and DPL (Google Maps API), etc. This causes our system to rely heavily on the internet. Once the service is suspended or there is no access to the internet, the system will not work. In the follow-up jobs, we must integrate offline solutions so that the use of navigation can be freer from environmental constraints.

The knowledge graph should provide more professional content for navigation applications including more relationship attributes, such as opening hours, allowable ages, or occasional restrictions. When the knowledge graph contains more information, the navigation location suggestions will be more helpful.

5. Conclusions

Understanding the user and responding in accordance with the given context is the main goal of the dialogue system. We have designed a method for using the knowledge graph as a knowledge base for reasoning dialogue to obtain the user's destination. The same method can be extended to confirm destination changes, indoor navigation, or route planning. Our system can become a good VUI to communicate with micro-navigation, wearable devices, or any smart device.

A traditional VUI uses commands to accomplish user's requirements. It converts speech into instructions. Users must make all decisions by themselves by speaking. We hope that the services of the system can be more user-friendly, so that it can guide users to gradually discover their needs. We believe this approach is closer to human thinking.

We have designed a dialogue system that integrates knowledge graphs and uses reasoning algorithm to guide users' destinations. We proposed a concrete and feasible macro-navigation architecture, and verified it in the real-world. After the experiment, we learned the importance of handling misunderstandings and the defects of over-reliance on cloud service. In addition, we also found that by improving the professionalism of the knowledge graph content, it will be very helpful for reasoning and achieve more accurate destination.

The main contribution of this paper is the design of a dialogue system architecture based on the knowledge graph which can complete the function of the dialogue system in DST. Additionally, another contribution is the use of the PRA algorithm to implement reasoning for navigation destinations.

Human natural language is widely used and all-encompassing. The general dialogue system is still difficult to meet the needs of casual chat, but the dialogue in a specific domain may be better. The semantic scope of the navigation system is very limited and concentrated, which is very suitable to becoming the best practice for dialogue in a specific domain.

In the future, we want to use the dialogue system with domain-related knowledge graphs in other fields such as medicine, law, or insurance. An intelligence dialogue is good for both the visually impaired people and any ordinary person.

Author Contributions: Conceptualization, C.-H.C. and M.-F.S.; methodology, M.-F.S.; software, M.-F.S.; validation, C.-H.C., M.-F.S. and S.-H.C.; formal analysis, C.-H.C.; investigation, S.-H.C.; resources, S.-H.C.; data curation, S.-H.C.; writing—original draft preparation, M.-F.S.; writing—review and editing, C.-H.C.; visualization, M.-F.S.; supervision, C.-H.C.; project administration, C.-H.C.; funding acquisition, C.-H.C. All authors have read and agreed to the published version of the manuscript.

Funding: This research is supported by MOST Taiwan (MOST-109-2221-E-008-055 and MOST-110-2634-F-008-005).

Institutional Review Board Statement: Not applicable.

Informed Consent Statement: Not applicable.

Acknowledgments: The financial support by MOST Taiwan and PAIR Labs contributions for this work are gratefully acknowledged.

Conflicts of Interest: The authors declare no conflict of interest.

References

1. Norgate, S.H. Accessibility of urban spaces for visually impaired pedestrians. *Proc. Inst. Civ. Eng. Munic. Eng.* **2012**, *165*, 231–237. [CrossRef]
2. Md, M.I.; Muhammad, S.S.; Kamal, Z.Z.; Md, M.A. Developing Walking Assistants for Visually Impaired People: A Review. *IEEE Sens. J.* **2019**, *8*, 2814.
3. Chen, C.H.; Wu, M.C.; Wang, C.C. Cloud-based Dialog Navigation Agent System for Service Robots. *Sens. Mater.* **2019**, *31*, 1871–1891. [CrossRef]
4. Bradley, N.A.; Dunlop, M.D. An Experimental Investigation into Wayfinding Directions for Visually Impaired People. *Pers. Ubiquitous Comput.* **2005**, *9*, 395–403. [CrossRef]
5. Strothotte, T.; Fritz, S.; Michel, R.; Raab, A.; Petrie, H.; Johnson, V.; Reichert, L.; Schalt, A. Development of Dialogue Systems for the Mobility Aid for Blind People: Initial Design and Usability Testing. In Proceedings of the Second Annual ACM Conference on Assistive Technologies, Vancouver, BC, Canada, 15 April 1996; pp. 139–144.
6. Fruchterman, J. Archenstoneı́s orientation tools: Atlas Speaks and Strider. In Proceedings of the Conference on Orientation and Navigation Systems for Blind Persons, Hatfield, UK, 1–2 February 1995.
7. Golledge, R.G.; Klatzky, R.L.; Loomis, J.M.; Speigle, J.; Tietz, J. A geographical information system for a GPS based personal guidance system. *Int. J. Geogr. Inf. Sci.* **1998**, *12*, 727–749. [CrossRef]
8. Nicholas, A.B.; Mark, D.D. *Investigating Context-Aware Clues to Assist Navigation for Visually Impaired People*; University of Strathclyde: Glasgow, Scotland, UK, 2002.
9. Shekhar, J.; Deepak, B.; Siddharth, P. HCI Guidelines for Designing Website for Blinds. *Int. J. Comput. Appl.* **2014**, *103*, 30.
10. Stina, O. *Designing Interfaces for the Visually Impaired: Contextual Information and Analysis of User Needs*; Umeå University: Umeå, Sweden, 2017; pp. 14–17.
11. McTear, M.; Zoraida, C.; David, G. *The Conversational Interface: Talking to Smart Devices*; Springer: New York, NY, USA, 2016; pp. 214–218.
12. Hongshen, C.; Xiaorui, L.; Dawei, Y.; Jiliang, T. *A Survey on Dialogue Systems: Recent Advances and New Frontiers*; Data Science and Engineering Lab, Michigan State University: East Lansing, MI, USA, 2017; p. 1.
13. Zhuosheng, Z.; Jiangtong, L.; Pengfei, Z.; Hai, Z.; Gongshen, L. *Modeling Multi-Turn Conversation with Deep Utterance Aggregation*; Shanghai Jiao Tong University: Shanghai, China, 2018; pp. 3740–3752.
14. Kornai, A.; Pullum, G.K. The X-Bar Theory of Phrase Structure. *Linguist. Soc. Am.* **1990**, *66*, 24–50.
15. Arthur, W.B. Inductive Reasoning and Bounded Rationality. *Am. Econ. Rev.* **1994**, *84*, 406–411.
16. Feng, N.; Ce, Z.; Christopher, R.; Jude, S. Elementary: Large-scale knowledge-base construction via machine learning and statistical inference. *Int. J. Semant. Web Info. Sys.* **2012**, *8*, 1–21.
17. Zhao, Y.J.; Li, Y.L.; Lin, M. A Review of the Research on Dialogue Management of Task-Oriented Systems. *J. Phys. Conf. Ser.* **2019**, *1267*, 12–25.
18. Google Cloud Platform. Available online: https://cloud.google.com/ (accessed on 1 March 2021).
19. Graham, K.; Jeremy, J.C.; Brian, M. RDF 1.1 Concepts and Abstract Syntax. W3C Recommendation 2014. Available online: https://www.w3.org/TR/rdf11-concepts/ (accessed on 1 March 2021).
20. Aprilia, W. English Passive Voice: An X-Bar Theory Analysis. *Indones. J. Engl. Lang. Stud.* **2018**, *4*, 69–75.
21. Cloud Natural Language API. Available online: https://cloud.google.com/natural-language/ (accessed on 1 March 2021).
22. Pearl, J.; Azaria, P. *Graphoids: A Graph-Based Logic for Reasoning about Relevance Relations*; University of California: Los Angeles, CA, USA, 1985; pp. 357–363.
23. Francesco, M.D.; Maurizio, L.; Daniele, N. *Reasoning in Description Logics. Principles of Knowledge Representation*; CSLI-Publications: Stanford, CA, USA, 1996; pp. 191–236.
24. Ni, L.; Tom, M.M.; William, W.C. Random Walk Inference and Learning in a Large Scale Knowledge Base. In Proceedings of the 2011 Conference on Empirical Methods in Natural Language Processing, Edinburgh, UK, 27–31 July 2011; pp. 529–539.
25. Antoine, B.; Nicolas, U.; Alberto, G.D. *Translating Embeddings for Modeling Multi-Relational Data*; Neural Information Processing Systems (NIPS): South Lake Tahoe, CA, USA, 2013; pp. 2787–2795.
26. Luis, A.G.; Christina, T.; Katja, H. Association Rule Mining under Incomplete Evidence in Ontological Knowledge Bases. In Proceedings of the 22nd International Conference on World Wide Web, Rio de Janeiro, Brazil, 13 May 2013; pp. 413–422.
27. Xiong, W.; Thien, H.; William, Y.W. DeepPath: A Reinforcement Learning Model for Knowledge Graph Reasoning. *arXiv* **2017**, arXiv:1707.06690.
28. Ni, L.; William, W.C. Relational retrieval using a combination of path-constrained random walks. *Mach. Learn.* **2010**, *81*, 53–67.
29. Google Maps Platform. Available online: https://cloud.google.com/maps-platform/ (accessed on 1 March 2021).
30. Apache Jena. Available online: https://jena.apache.org/ (accessed on 1 March 2021).

Article

Human Pose Detection for Robotic-Assisted and Rehabilitation Environments

Óscar G. Hernández, Vicente Morell, José L. Ramon and Carlos A. Jara *

Human Robotics Group, University of Alicante, San Vicente del Raspeig, 03690 Alicante, Spain; oghernandez@unah.edu.hn (Ó.G.H.); vicente.morell@ua.es (V.M.); jl.ramon@ua.es (J.L.R.)
* Correspondence: carlos.jara@ua.es; Tel.: +34-965903400 (ext. 1094)

Featured Application: The proposed technology is useful for the estimation of human biomarkers in rehabilitation processes or for any application that needs human pose estimation.

Abstract: Assistance and rehabilitation robotic platforms must have precise sensory systems for human–robot interaction. Therefore, human pose estimation is a current topic of research, especially for the safety of human–robot collaboration and the evaluation of human biomarkers. Within this field of research, the evaluation of the low-cost marker-less human pose estimators of OpenPose and Detectron 2 has received much attention for their diversity of applications, such as surveillance, sports, videogames, and assessment in human motor rehabilitation. This work aimed to evaluate and compare the angles in the elbow and shoulder joints estimated by OpenPose and Detectron 2 during four typical upper-limb rehabilitation exercises: elbow side flexion, elbow flexion, shoulder extension, and shoulder abduction. A setup of two Kinect 2 RGBD cameras was used to obtain the ground truth of the joint and skeleton estimations during the different exercises. Finally, we provided a numerical comparison (RMSE and MAE) among the angle measurements obtained with OpenPose, Detectron 2, and the ground truth. The results showed how OpenPose outperforms Detectron 2 in these types of applications.

Keywords: human–robot interaction; human pose estimation; robotic rehabilitation

Citation: Hernández, Ó.G.; Morell, V.; Ramon, J.L.; Jara, C.A. Human Pose Detection for Robotic-Assisted and Rehabilitation Environments. *Appl. Sci.* **2021**, *11*, 4183. https://doi.org/10.3390/app11094183

Academic Editors: Alessandro Di Nuovo and Manuel Armada

Received: 26 March 2021
Accepted: 30 April 2021
Published: 4 May 2021

Publisher's Note: MDPI stays neutral with regard to jurisdictional claims in published maps and institutional affiliations.

1. Introduction

A high demand of services for assisted and rehabilitation environments is expected from the health status of the world due to the COVID-19 pandemic. Currently, according to the WHO (World Health Organization), existing rehabilitation services have been disrupted in 60–70% of countries due to this pandemic in order to avoid human contact. Therefore, countries must face major challenges to ensure the health of their population. Robotic platforms are a great solution to ensure assistance and rehabilitation for disabled people using human–robot interaction (HRI) capabilities. HRI is currently a topic of research that contributes by means of several research approaches for the physical and/or social interaction of humans and robotic systems [1] in order to achieve a goal together.

The number of people with congenital and/or acquired disabilities are quickly increasing, and therefore, there are many dependents who lack the necessary autonomy for a fully independent life. Among them, stroke is one of the main causes of these acquired disabilities throughout the world. Acquired brain injury (ABI) is a clinical-functional situation triggered by an injury of any origin that acutely affects the brain, causing neurological deterioration, functional loss, and poor quality of life as a result. It can be due to various causes, with stroke and head trauma the most frequent in our environment. Patients with ABI suffer cognitive and motor sequelae. In stroke patients, motor sequelae are usually more severe in the upper limb. In published studies, it has been reported that 30–60% of hemiplegic patients due to a stroke remain with severely affected upper limbs after 6 months of the event, and only 5–20% manage a complete functional recovery.

Physical medicine and rehabilitation are the most important treatment methods in ABI because they help patients reutilize their limbs at maximum capacity. Intensive therapies and repetitive task-based exercises are very effective treatments for motor skills recovery [2]. One of the most important processes of physical therapy requires manual exercises, in which the physiotherapist and the patient must have one-to-one interaction. The goal of the physiotherapist in this process is to help the patients achieve a normal standard of range of motion in their limbs and to strengthen their muscles. Rehabilitation robotic platforms pursue the recovery of impaired motor function. The majority of rehabilitation robotics research to date has focused on passive post-stroke exercises (e.g., [3,4]). The use of assistive robotics in rehabilitation allows the assistance of the physiotherapist in certain exercises that require repeated movements with high precision. The robot can fulfil the requirements of the cyclic movements in rehabilitation. Additionally, robots can successfully control the forces applied and can monitor the therapy results objectively by using their sensors.

Robots intended for upper limb rehabilitation can accomplish active and passive bilateral and unilateral motor skills training for the wrist, forearm, and shoulder. MIT-Manus is one of the most well-known upper limb rehabilitation robots [5]. It was developed for unilateral shoulder or elbow rehabilitation. MIME is another well-known upper limb rehabilitation robot, developed for elbow rehabilitation using the master–slave concept [6]. The movement of the master side of the robot is reproduced on the slave side. The 2-DOF robot can perform flexion–extension and pronation–supination movements. The Assistive Rehabilitation and Measurement (ARM) Guide is a bilateral rehabilitation system for upper limb rehabilitation using an industrial robot [7]. It assists the patient in following a trajectory. It also serves as a basis for the evaluation of several key motor impairments, including abnormal tone, incoordination, and weakness. The GENTLE/s system uses a haptic interface and virtual reality techniques for rehabilitation. The patients can move their limbs in a three-dimensional space with the aid of the robot [8]. The authors of [9] presented a rehabilitation robot with minimum degrees of freedom to train the arm in reaching and manipulation, called reachMAN2. All these previous robotic devices provide the potential for patients to carry out more exercise with limited assistance, and dedicated robotic devices can progressively adapt to the patients' abilities and quantify the improvement of the subject.

Robotic platforms for assistance and rehabilitation must have precise sensory systems for HRI. Therefore, they must recognize human pose or human gestures to improve the performance and safety of human–robot collaboration in these environments [10,11]. Our study sought to obtain a marker-less good pose estimation using a low-cost RGB camera for upper limb robotic rehabilitation environments. We set-up a multiple-RGBD-calibrated-cameras system to measure the goodness of the available methods.

2. Human Pose Detection and Body Feature Extraction: A State of the Art

The human body is a very complex system composed of many limbs and joints, and the exact detection of the position of the joints in 2D or 3D is a challenging task [12], as it requires a specific assumption within biomechanics research in robotic rehabilitation environments [13]. In addition, HRI environments are complex and nondeterministic, and it is not easy to ensure the user's safety during interaction with the robot. Currently, this assumption is a research topic in other areas, such as Industry 4.0 [14,15]. The resolution of this issue involves constant position tracking, intention estimation, and action prediction of the user. This problem can be faced by a proper sensory system. On the one hand, some contributions employ inertial measurement units (IMUs) for motion capture, especially in medical applications and motor rehabilitation analysis [16,17]. However, this type of sensor requires the correct placement of passive/active markers on the body before each capture session, and they are insufficient for HRI environments.

On the other hand, this issue can also be faced as a computer vision problem, basically using two approaches: marker-based and marker-less. Marker-based approaches, such as motion capture systems (MoCap), have significant environmental constraints (markers in the human body) and are relatively complex, expensive, and difficult to maintain. Marker-less approaches have fewer environmental constraints, and they can give a new understanding about human movements [18]. This issue requires the processing of complex information to develop an algorithm that recognizes human poses or skeletons from images. Therefore, an easy-to-use marker-less motion capture method is desirable for these robotic rehabilitation environments. In this paper, we analyzed the performance of the estimation of shoulder and elbow angles for the development of rehabilitation exercises using CNN (convolutional neural network)-based human pose estimation methods.

There is extensive research about marker-less approaches for human tracking motion. In these approaches, depth cameras such as Kinect (RGB-D) provide additional information about the 3D shape of the scene. Kinect has become an important 3D sensor, and it has received a lot of attention thanks to its rapid human pose recognition system. Its low-cost, reliability, and rapid measurement characteristics have made the Kinect the primary 3D measuring device for indoor robotics, 3D scene reconstruction, and object recognition [19]. Several approaches for real-time human pose recognition have been performed only using a single sensor [20–22], but it can have substantial errors with partial occlusions.

In recent years, the use of deep learning techniques for 3D human pose estimation has become a common approach in HRI systems. These computer vision techniques usually train a neural network from labeled images in order to estimate human pose. As a reference, some research works obtain the 3D pose estimation using a single-view RGB camera image or multi-view camera images [23,24]. These accurate methods encounter fewer problems regarding the cameras' position and calibration issues in comparison to RGB-D approaches. However, 3D human pose detection for assisted and rehabilitation robotic environments needs further improvements to achieve real-time tracking for human motion analysis with enough accuracy [25].

In our comparison, we decided to use the Kinect v2 sensor, which has a high accuracy in joint estimation while providing skeletal tracking, as described in [26]. Additionally, some research about this fact has been presented in [27], where the validity of Kinect v2 for clinical motion was compared with a MoCap system, and the error was reported to be less than 5%. We employed the skeleton obtained by two Kinect sensors as our ground truth to measure the performance of the estimation of shoulder and elbow angles using two CNN (convolutional neural network)-based human pose estimation methods in rehabilitation exercises. The selected CNN-based methods were OpenPose [28] and Detectron2 [29]. OpenPose is a multi-person pose detection system, and it can detect a total of 135 body points from a digital image [28,30]. OpenPose has been trained to produce three distinct pose models. They differ from one another in the number of estimated key points: (a) MPII is the most basic model and can estimate a total of 15 important key points: ankles, knees, hips, shoulders, elbows, wrists, necks, torsos, and head tops. (b) The COCO model is a collection of 18 points including some facial key points. (c) BODY pose provides 25 points consisting of COCO + feet keypoints [30,31]. Detectron2 was built by Facebook AI Research (FAIR) to support the rapid implementation and evaluation of novel computer vision research. Detectron2 is a ground-up rewrite of the previous Detectron version and it originates from the Mask R-CNN benchmark. Detectron2 includes high-quality implementations of state-of-the-art object detection algorithms, including DensePose [29].

3. Materials and Methods

The architecture of the vision system is composed of two RGBD cameras (Microsoft Kinect Xbox One, also known as Kinect v2, Microsoft, Albuquerque, NM, USA) and a webcam connected to a computer network. Each Kinect is connected to a client computer, which estimates the user skeleton joint tracking through the Microsoft Kinect Software Development Kit (SDK). Microsoft released the Kinect sensor V2 in 2013, which incorporates a RGB camera with a resolution of 1920×1080 pixels and a depth sensor with a resolution of 512×424 pixels and a working range of 5–450 cm, a $70° \times 60°$ field of view, and a frame rate of 15–30 fps. Data from the sensor can be accessed using the Kinect for Windows SDK 2.0, which allows tracking up to 6 users simultaneously, each with 25 joints. For each joint, the three-dimensional position is provided, as well as the orientation as quaternions. The center of the IR camera lens represents the origin of the 3D coordinate system [32,33].

The Microsoft SDK was designed for Windows platforms; therefore, rosserial is used to communicate between Windows platforms and Linux [34]. Three PCs are used for the system architecture. One of them works both as a client and server. The detailed hardware description is shown in Table 1. The RGB webcam is connected to a client PC equipped with a graphics card (GPU). Both the webcam and GPU are used by the OpenPose and Detectron2 methods for human pose estimation. The overall system topology is shown in Figure 1.

Table 1. Hardware setup of our system.

	Server-Client	Client 2	Client 3
OS	Ubuntu 18.04.03 Desktop (64 bit), (Canonical, London, UK)	Windows 10 Pro (64 bit), (Microsoft, Albuquerque, NM, USA)	Windows 10 Pro (64 bit)
Processor	Intel® Core™ i7-9750, (Intel, Santa Clara, CA, USA)	Intel® Core™ i5-8250U	Intel® Core™ i7-4700MQ
Memory	16 GB	16 GB	16 GB
GPU	NVIDIA GeForce GTX 1650 GDDR5 @4 GB (128 bits), (NVIDIA, Santa Clara, CA, USA)		

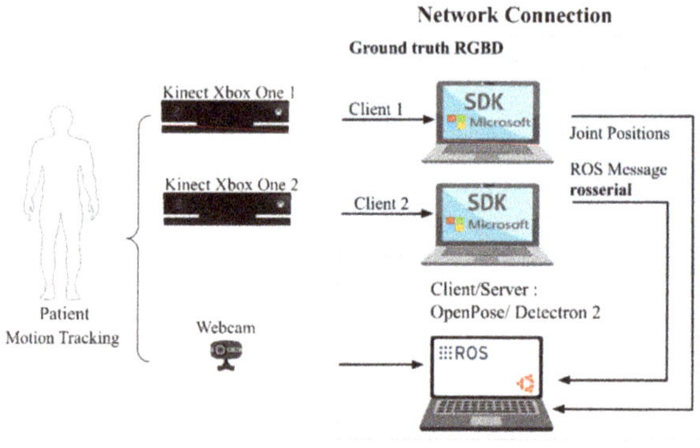

Figure 1. An overview of the proposed system.

Cameras Calibration

In order to compare the different pose estimation methods, all 3 cameras need to be calibrated in a common coordinate system. Calibration of the cameras was performed using the OpenCV multiple cameras calibration package [35]. A checkerboard pattern with known dimensions is shown so that at least two cameras can identify it at the same time. To obtain the ground truth, information of the extrinsic parameters of the cameras (translation and rotation matrix) is required, then a 3D-to-2D projection must be made in the image plane to be able to compare with the information provided from OpenPose or Detectron 2 (see Equations (1) and (2)). The acquisition and processing scheme of the data is shown in Figure 2.

$$\begin{bmatrix} X_{CK1} \\ Y_{CK1} \\ Z_{CK1} \end{bmatrix} = R \begin{bmatrix} X_{K1} \\ Y_{K1} \\ Z_{K1} \end{bmatrix} + T \tag{1}$$

$$u = \left(\frac{x}{z}\right) * f_x + c_x v = \left(\frac{y}{z}\right) * f_y + c_y \tag{2}$$

where (X_{K1}, Y_{K1}, Z_{K1}) are the coordinates of a 3D point in the coordinate system of the Kinect 1 (IR camera), $(Xc_{K1}, Yc_{K1}, Zc_{K1})$ are the coordinates of a 3D point calculated from the Kinect 1, T is the transfer vector, R is the rotation matrix, (u, v) are the coordinates of the projection point in pixels, (cx, cy) is the image center (IR camera), and (fx, fy) are the focal lengths expressed in pixel units (IR camera).

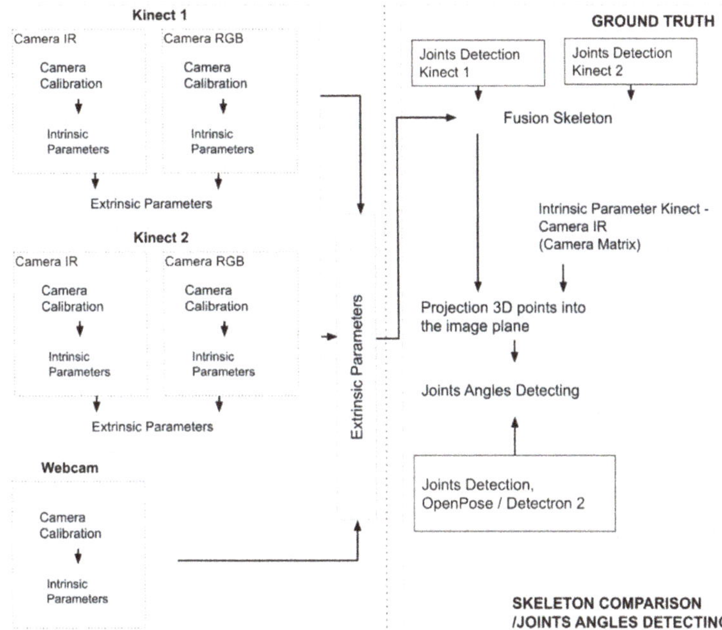

Figure 2. Data acquisition and processing scheme.

4. Experimental Setup

4.1. Cameras Position

The two Kinect sensors were located orthogonally to each other, as described in Figure 3a. This distribution allows the elimination of the problem of data loss caused by self-occlusion [36,37]. The distribution of the laboratory hardware setup is shown in Figure 3b. With this configuration, more precise data can be obtained on rehabilitation exercises that focus on the limbs. Finally, the webcam was located just above the Kinect 2 to reduce the errors in extrinsic parameters and to obtain a similar view with the Kinect 2.

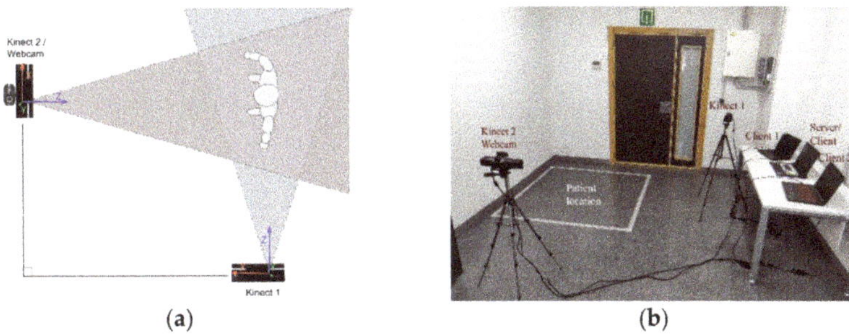

Figure 3. Our system: (**a**) Cameras distribution, (**b**) Working zone.

4.2. Joint Angle Measurement

The joint angle was measured as the relative angle between the longitudinal axis of two adjacent segments. These segments were composed of three points in the 2D space: a starting point, a middle point, and an end point. For the elbow joint angle, the adjacent segments were the upper arm and the forearm, respectively. Figure 4 shows the elbow and shoulder joint angles measured in this study. Let u and v be vectors representing two adjacent segments, where the angle between u and v is equal to:

$$\theta = \cos^{-1}\left(\frac{u \cdot v}{|u||v|}\right) \tag{3}$$

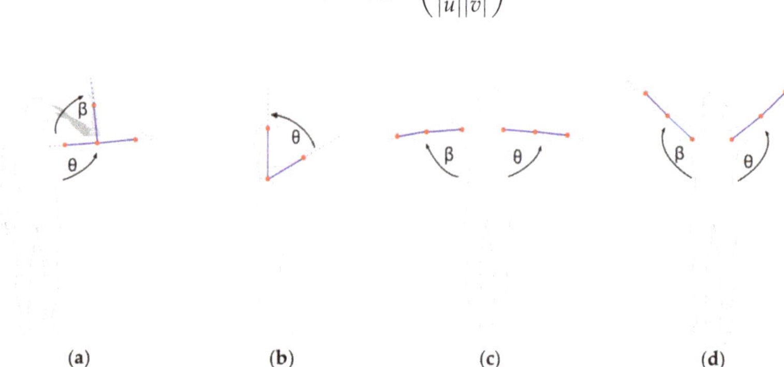

(a)   (b)   (c)   (d)

Figure 4. Rehabilitation exercises: (**a**) elbow side flexion, (**b**) elbow flexion, (**c**) shoulder extension, and (**d**) shoulder abduction.

4.3. Rehabilitation Exercises

Four upper limb rehabilitation exercises were proposed: elbow side flexion, elbow flexion, shoulder extension, and shoulder abduction (Figure 4). During the execution of the exercises, the cameras capture the information of the desired joints of the patient pose, and this information is used to calculate the angles θ and β obtained by the different systems.

4.4. Ground Truth

The ground truth of the pose estimation was calculated using the skeletons provided by the two Kinect cameras. As previously mentioned, these cameras were located at approximately 90° from each other to obtain accurate data on rehabilitation exercises that focus on the upper limb where one camera does not give fully reliable estimations. Figure 5 shows four examples of 3D human pose estimations obtained by the Kinect 2 (blue color), and the skeleton fusions (red color) during an elbow side flexion exercise. The skeleton fusion was calculated with a simple average of the Kinect 2 skeleton and the projected skeleton from the Kinect 1, which was calculated using the information obtained in the calibration phase and using Equation (3). When performing this projection, a difference is expected between the coordinates of the Kinect 2 (main camera, front view) and the Kinect 1 (auxiliary camera, side view). This difference is due to the viewing angle of each Kinect and the volume of the joints of the human.

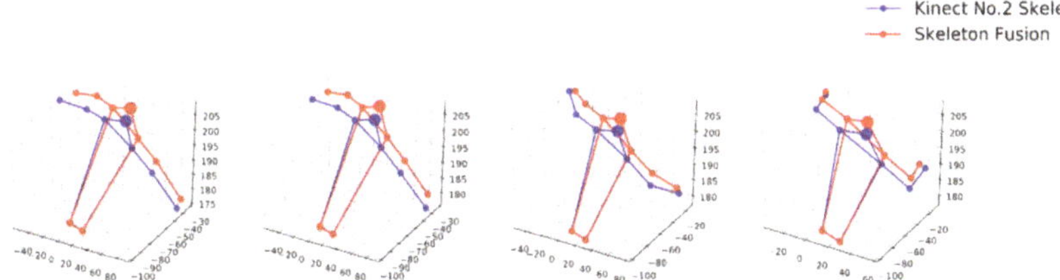

Figure 5. 3D skeleton poses obtained by Kinect 2 and skeleton fusion.

Figure 6 shows the X, Y, and Z positions of the left and right wrist of the two Kinects (Kinect 1 and Kinect 2) during a shoulder extension exercise and the projected points (Projected 1). As stated before, an error was expected in the calculation of the fused skeleton, and the results show how even with the fully calibrated system, we obtained some errors. For the left wrist, we measured MAEs (mean absolute errors) of 7.35, 2.16, and 3.71 cm for the coordinates X, Y, and Z, respectively. For the right wrist, we measured MAEs of 7.70, 2.90, and 3.25 cm for the coordinates X, Y, Z, respectively.

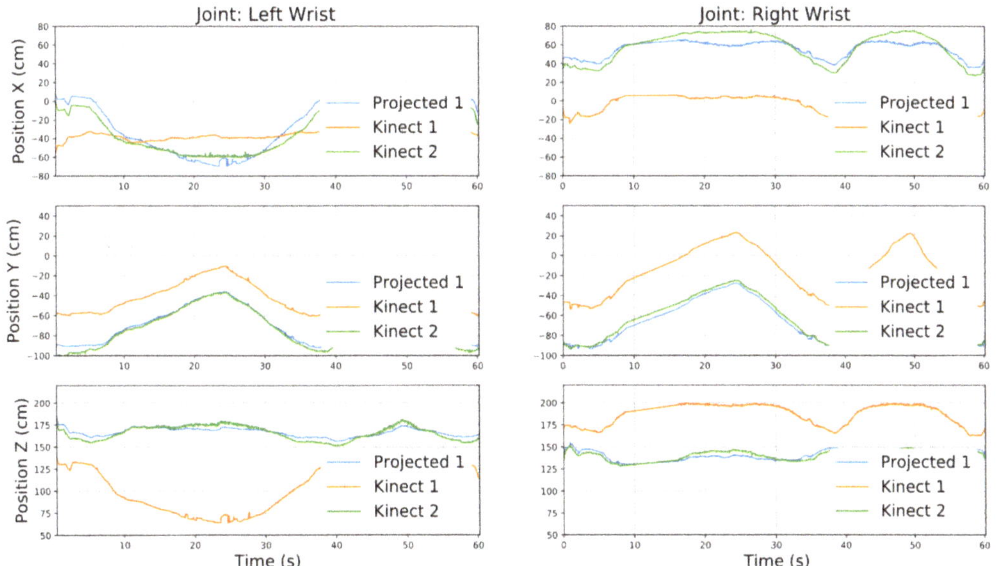

Figure 6. X, Y, and Z positions of left and right wrist by Kinect 1, Kinect 2, and projected position.

5. Experiments

The following experiments show the precision of the angles calculated using the OpenPose and Detectron 2 approaches. Only the necessary angles are shown in each experiment. The movements of exercises 1 and 2 involved only one angle, while exercises 3 and 4 involved the two aforementioned elbow and shoulder angles. A video of the experiments is available online [38]. A summary of the results of all experiments is shown in Section 6 (Results section).

5.1. Exercise 1: Elbow Side Flexion

The first experiment was the elbow side flexion exercise. Figure 7a shows the left elbow angle calculated with both approaches for all positions in the exercise. The left elbow angle varied between 50° and 180° with OpenPose and between 40° and 180° with

Detectron 2. For OpenPose, the root-mean-square error (RMSE) was 9.23° and the mean absolute error (MAE) was 7.53°, while for Detectron 2, the RMSE was 13.64° and the MAE was 9.32°.

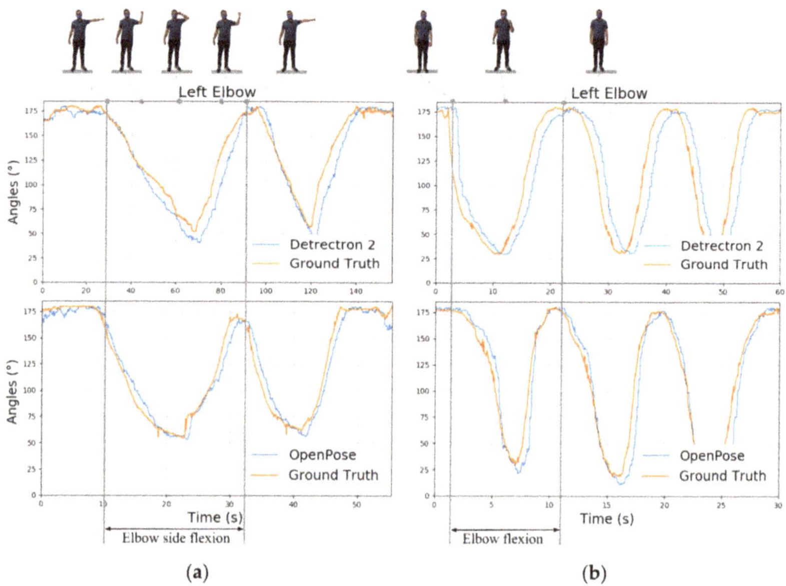

Figure 7. Exercise 1: elbow side flexion angles for two iterations (**a**); exercise 2: elbow flexion for three iterations (**b**).

5.2. Exercise 2: Elbow Flexion

The second experiment was the left elbow flexion exercise where the angle of view of the main points was not the most favorable. Figure 7b shows the results of the calculation of angles. The left elbow angle varied between 11 and 180° (i.e., straight arm) with OpenPose and between 30° and 180° with Detectron 2. For OpenPose, RMSE = 15.84° and MAE = 9.74°, while for Detectron 2, RMSE = 27.27° and MAE = 20.03°.

5.3. Exercise 3: Shoulder Extension

The third and fourth experiments aimed to measure the precision of both approaches when performing rehabilitation exercises with both arms in shoulder extension and abduction. Figure 8 shows the angles for the left and right shoulder in the shoulders' extension exercise. The left shoulder angle varied between 5° and 91° with OpenPose and between 4° and 98° with Detectron 2. The right shoulder angle varied between 7° and 102° with OpenPose and between 3° and 95° with Detectron 2. For the left shoulder, RMSE = 3.57° and MAE = 2.96° for OpenPose, while RMSE = 7.99° and MAE = 7.15° for Detectron 2. For the right shoulder, RMSE = 8.67° and MAE = 8.16° for OpenPose, while RMSE = 11.70° and MAE = 9.24° for Detectron 2.

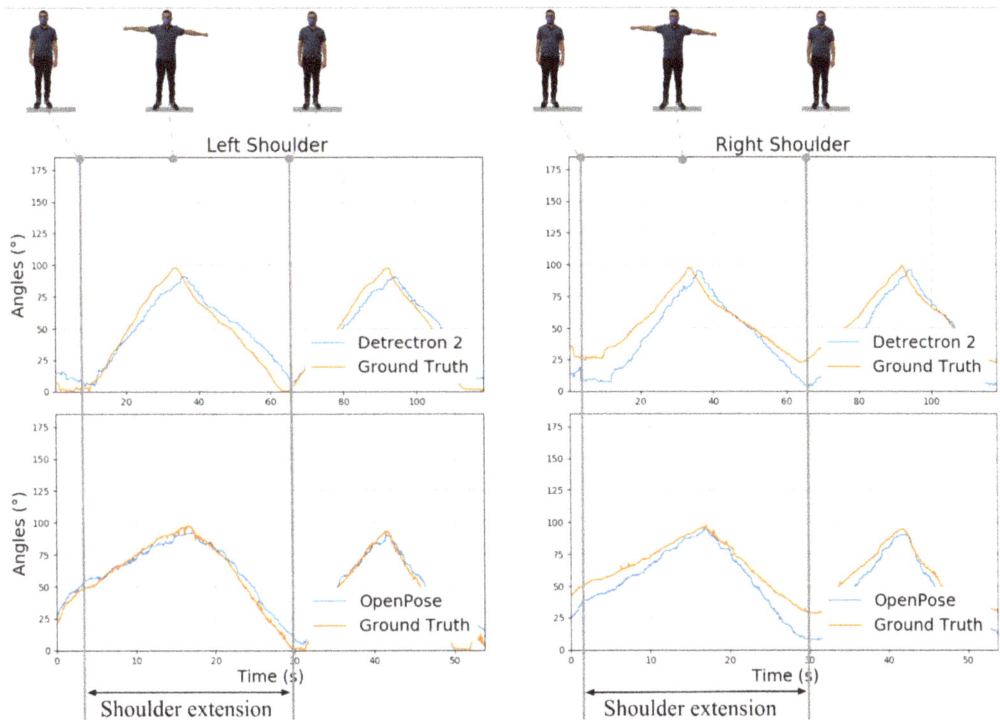

Figure 8. Exercise 3: angles of both shoulders in extension rehabilitation exercise for two iterations.

5.4. Exercise 4: Shoulder Abduction

Figure 9 shows the angles for the left shoulder and right shoulder in the shoulders' abduction exercise. The left shoulder angle varied between 80° and 128° with OpenPose and between 79° and 127° with Detectron 2. The right shoulder angle varied between 85° and 132° with OpenPose and between 80° and 138° with Detectron 2. For the left shoulder angle, RMSE = 4.68° and MAE = 4.17° for OpenPose, while RMSE = 11.79° and MAE = 7.89° for Detectron 2. For the right shoulder angle, RMSE = 4.16° and MAE = 3.74° for OpenPose, while RMSE = 12.74° and MAE = 9.03° for Detectron 2.

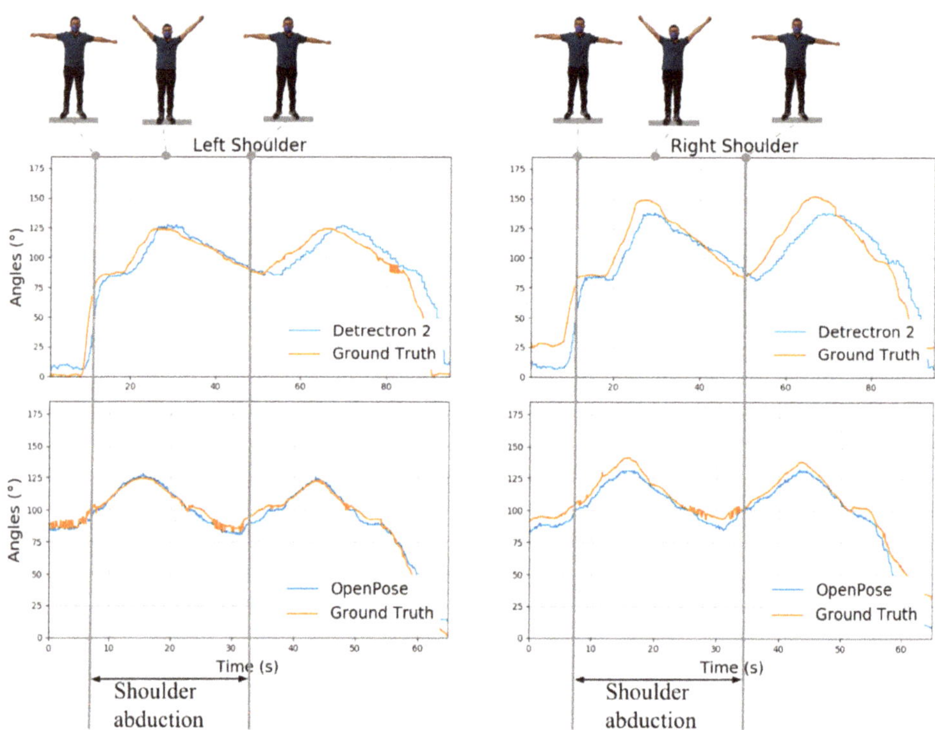

Figure 9. Exercise 4: Angles of both shoulders in shoulder abduction rehabilitation exercise for two iterations.

6. Results

Figure 10 shows a comparison of the RMSEs (root-mean-square errors) of OpenPose and Detectron 2 for each rehabilitation angle calculated in the exercises. OpenPose obtained a lower RMSE than Detectron 2 did for all four proposed rehabilitation exercises. For the exercises where the viewing angle of the webcam was not favorable, a large RMSE was obtained for both approaches. An example of these errors can be seen in the flexion left elbow exercise, which had RMSEs of 15.84° with OpenPose and 27.27° with Detectron 2. The best results were obtained by OpenPose in exercises 3 and 4 where it was easier to estimate the movement of the arms than to obtain an estimated RMSE below 5°.

To visualize the error during each exercise step, we calculated the absolute error of each method. Figure 11 shows the absolute error for each rehabilitation exercise according to both libraries. We can see again that during the exercise of flexion of the left elbow (row b), both libraries achieved high absolute errors. The results show how OpenPose had fewer error peaks, and it seemed more stable for most of the angles checked.

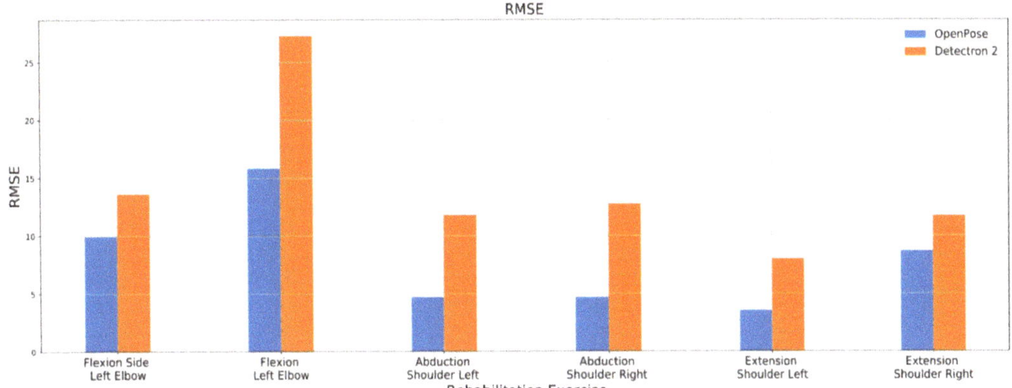

Figure 10. RMSE of the four rehabilitation exercises compared to ground truth.

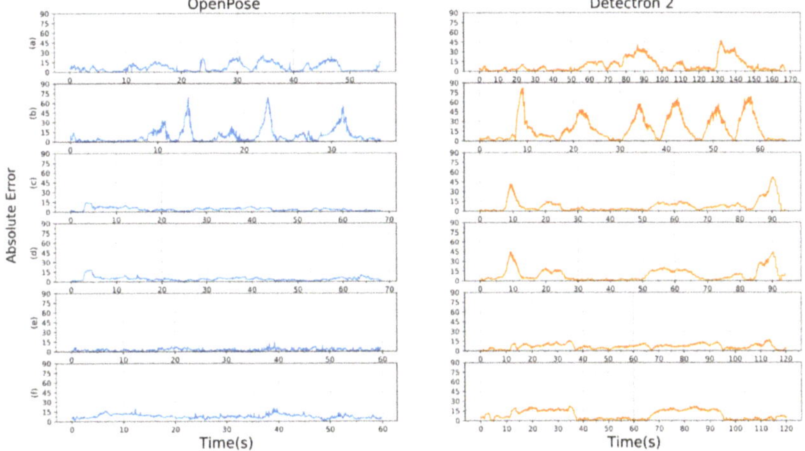

Figure 11. Absolute error for the four rehabilitation exercises compared to ground truth: (**a**) flexion side left elbow, (**b**) flexion left elbow, (**c**) abduction left shoulder, (**d**) abduction right shoulder, (**e**) extension left shoulder, and (**f**) extension right shoulder.

7. Discussion

In this article, we compared the performance of estimating shoulder and elbow angles for rehabilitation exercises using CNN-based human pose estimation methods: Open-Pose and Detectron2. Qualitatively, for the four proposed rehabilitation exercises, better results were obtained with OpenPose. OpenPose had an average RMSE of 7.9° and Detectron2 had an RMSE of 14.18°. However, for the elbow flexion exercise, which had a worse angle of view, both methods obtained high errors. According to [39], evaluation therapists tend to underestimate the range of motion by 9.41° on average for any joint movement of the upper limb. Therefore, with the results obtained in this approach, it can be concluded that OpenPose is an adequate library for evaluating patient performance in rehabilitation programs that involve the following exercises: left elbow side flexion, shoulder abduction, and shoulder extension.

Regarding the response time of the analyzed pose estimators, the performance of these methods is related directly to the available GPU. In our study, we measured a performance between 6.7 and 13 FPS with OpenPose and between 1.8 and 3 FPS with Detectron 2.

The kind of exercises related to upper limb rehabilitation is smooth and relatively slow, so the performance of OpenPose is high enough to monitor the exercises and provide useful information for rehabilitation therapy.

The major limitations of the present study were mainly related to the ground truth used; many approaches use three-dimensional motion analysis devices such as the VICON motion system or Optotrak. However, the equipment is expensive, and it requires a conditioned environment and technical skills for attaching sensors. We decided to use Kinect 2 as our ground truth because of the cost and because this sensor has a high accuracy in joint estimation while providing skeletal tracking. Another limitation of the study was that the system was only tested on a single healthy subject who participated in a single experimental session. A study with a larger group of subjects and different positions should be examined to compare the quality of the estimate of the human pose by both methods. In future works, we intend to collect data from more participants and extend this work to lower-limb movement estimations.

Author Contributions: Conceptualization, C.A.J., V.M. and Ó.G.H.; methodology, C.A.J., V.M. and J.L.R.; software, Ó.G.H.; validation, C.A.J., V.M., J.L.R. and Ó.G.H.; investigation, C.A.J., V.M. and Ó.G.H.; resources, C.A.J.; writing—original draft preparation, C.A.J., V.M., J.L.R. and Ó.G.H.; writing—review and editing, C.A.J., V.M., J.L.R. and Ó.G.H. All authors have read and agreed to the published version of the manuscript.

Funding: Óscar G. Hernández holds a grant from the Spanish Fundación Carolina, the University of Alicante, and the National Autonomous University of Honduras.

Institutional Review Board Statement: Not applicable.

Informed Consent Statement: Informed consent was obtained from all subjects involved in the study.

Data Availability Statement: Not applicable.

Conflicts of Interest: The authors declare no conflict of interest.

References

1. Olaronke, I.; Oluwaseun, O.; Rhoda, I. State of The Art: A Study of Human-Robot Interaction in Healthcare. *Int. J. Inf. Eng. Electron. Bus.* **2017**, *9*, 43–55. [CrossRef]
2. Claflin, E.S.; Krishnan, C.; Khot, S.P. Emerging Treatments for Motor Rehabilitation after Stroke. *Neurohospitalist* **2015**, *5*, 77–88. [CrossRef] [PubMed]
3. Krebs, H.I.; Volpe, B.T.; Ferraro, M.; Fasoli, S.; Palazzolo, J.; Rohrer, B.; Edelstein, L.; Hogan, N. Robot-Aided Neuro-Rehabilitation: From Evidence-Based to Science-Based Rehabilitation. In *Topics in Stroke Rehabilitation*; Thomas Land Publishers Inc.: St. Louis, MO, USA, 2002; pp. 54–70. [CrossRef]
4. Burgar, C.G.; Lum, P.S.; Shor, P.C.; Van Der Loos, H.F.M. Development of Robots for Rehabilitation Therapy: The Palo Alto VA/Stanford Experience. *J. Rehabil. Res. Dev.* **2000**, *37*, 663–673. [PubMed]
5. Igo Krebs, H.; Hogan, N.; Aisen, M.L.; Volpe, B.T. Robot-Aided Neurorehabilitation. *IEEE Trans. Rehabil. Eng.* **1998**, *6*, 75–87. [CrossRef] [PubMed]
6. Lum, P.S.; Burgar, C.G.; Shor, P.C.; Majmundar, M.; Van der Loos, M. Robot-Assisted Movement Training Compared with Conventional Therapy Techniques for the Rehabilitation of Upper-Limb Motor Function after Stroke. *Arch. Phys. Med. Rehabil.* **2002**, *83*, 952–959. [CrossRef] [PubMed]
7. Reinkensmeyer, D.J.; Kahn, L.E.; Averbuch, M.; McKenna-Cole, A.; Schmit, B.D.; Zev Rymer, W. Understanding and Treating Arm Movement Impairment after Chronic Brain Injury: Progress with the ARM Guide. *J. Rehabil. Res. Dev.* **2000**, *37*, 653–662. [PubMed]
8. Richardson, R.; Brown, M.; Bhakta, B.; Levesley, M.C. Design and Control of a Three Degree of Freedom Pneumatic Physiotherapy Robot. *Robotica* **2003**, *21*, 589–604. [CrossRef]
9. Zhu, T.L.; Klein, J.; Dual, S.A.; Leong, T.C.; Burdet, E. ReachMAN2: A Compact Rehabilitation Robot to Train Reaching and Manipulation. In Proceedings of the IEEE International Conference on Intelligent Robots and Systems, Chicago, IL, USA, 14–18 September 2014; pp. 2107–2113. [CrossRef]
10. Lioulemes, A.; Theofanidis, M.; Kanal, V.; Tsiakas, K.; Abujelala, M.; Collander, C.; Townsend, W.B.; Boisselle, A.; Makedon, F. MAGNI Dynamics: A Vision-Based Kinematic and Dynamic Upper-Limb Model for Intelligent Robotic Rehabilitation. *Int. J. Biomed. Biol. Eng.* **2017**, *11*, 158–167.
11. Park, D.; Kim, H.; Kemp, C.C. Multimodal Anomaly Detection for Assistive Robots. *Auton. Robot.* **2019**, *43*, 611–629. [CrossRef]

12. Sarafianos, N.; Boteanu, B.; Ionescu, B.; Kakadiaris, I.A. 3D Human Pose Estimation: A Review of the Literature and Analysis of Covariates. *Comput. Vis. Image Underst.* **2016**, *152*, 1–20. [CrossRef]
13. Lioulemes, A.; Theofanidis, M.; Makedon, F. Quantitative Analysis of the Human Upper-Limp Kinematic Model for Robot-Based Rehabilitation Applications. In Proceedings of the IEEE International Conference on Automation Science and Engineering, Fort Worth, TX, USA, 21–24 August 2016; pp. 1061–1066. [CrossRef]
14. Di Nardo, M. Developing a Conceptual Framework Model of Industry 4.0 for Industrial Management. *Ind. Eng. Manag. Syst.* **2020**, *19*, 551–560. [CrossRef]
15. Brito, T.; Lima, J.; Costa, P.; Matellán, V.; Braun, J. Collision Avoidance System with Obstacles and Humans to Collaborative Robots Arms Based on RGB-D Data. In *Advances in Intelligent Systems and Computing*; Springer: Berlin/Heidelberg, Germany, 2020; Volume 1092 AISC, pp. 331–342. [CrossRef]
16. Leardini, A.; Lullini, G.; Giannini, S.; Berti, L.; Ortolani, M.; Caravaggi, P. Validation of the Angular Measurements of a New Inertial-Measurement-Unit Based Rehabilitation System: Comparison with State-of-the-Art Gait Analysis. *J. Neuroeng. Rehabil.* **2014**, *11*, 136. [CrossRef] [PubMed]
17. Home-Xsens 3D Motion Tracking. Available online: https://www.xsens.com/ (accessed on 13 April 2021).
18. Mündermann, L.; Corazza, S.; Andriacchi, T.P. The Evolution of Methods for the Capture of Human Movement Leading to Markerless Motion Capture for Biomechanical Applications. *J. Neuroeng. Rehabil.* **2006**, *3*, 6. [CrossRef] [PubMed]
19. Smisek, J.; Jancosek, M.; Pajdla, T. 3D with Kinect. In *Consumer Depth Cameras for Computer Vision*; Springer: London, UK, 2013; pp. 3–25. [CrossRef]
20. Shotton, J.; Fitzgibbon, A.; Cook, M.; Sharp, T.; Finocchio, M.; Moore, R.; Kipman, A.; Blake, A. Real-Time Human Pose Recognition in Parts from Single Depth Images. In Proceedings of the IEEE Computer Society Conference on Computer Vision and Pattern Recognition, Colorado Springs, CO, USA, 20–25 June 2011; pp. 1297–1304. [CrossRef]
21. Ganapathi, V.; Plagemann, C.; Koller, D.; Thrun, S. Real Time Motion Capture Using a Single Time-of-Flight Camera. In Proceedings of the IEEE Computer Society Conference on Computer Vision and Pattern Recognition, San Francisco, CA, USA, 13–18 June 2010; pp. 755–762. [CrossRef]
22. Baak, A.; Müller, M.; Bharaj, G.; Seidel, H.-P.; Theobalt, C. *A Data-Driven Approach for Real-Time Full Body Pose Reconstruction from a Depth Camera*; Springer: London, UK, 2013; pp. 71–98. [CrossRef]
23. Pavllo, D.; Zürich, E.; Feichtenhofer, C.; Grangier, D.; Brain, G.; Auli, M. 3D Human Pose Estimation in Video with Temporal Convolutions and Semi-Supervised Training. In Proceedings of the 2019 Conference on Computer Vision and Pattern Recognition, Long Beach, CA, USA, 16–20 June 2019.
24. Moon, G.; Chang, J.Y.; Lee, K.M. Camera Distance-Aware Top-down Approach for 3D Multi-Person Pose Estimation from a Single RGB Image. In Proceedings of the International Conference on Computer Vision, Seoul, Korea, 27 October–2 November 2019.
25. Seethapathi, N.; Wang, S.; Saluja, R.; Blohm, G.; Kording, K.P. Movement Science Needs Different Pose Tracking Algorithms. *arXiv* **2019**, arXiv:1907.10226.
26. Wang, Q.; Kurillo, G.; Ofli, F.; Bajcsy, R. Evaluation of Pose Tracking Accuracy in the First and Second Generations of Microsoft Kinect. In Proceedings of the 2015 IEEE International Conference on Healthcare Informatics, ICHI 2015, Dallas, TX, USA, 21–23 October 2015; pp. 380–389. [CrossRef]
27. Ma, M.; Proffitt, R.; Skubic, M. Validation of a Kinect V2 Based Rehabilitation Game. *PLoS ONE* **2018**, *13*, e0202338. [CrossRef]
28. GitHub-CMU-Perceptual-Computing-Lab/Openpose: OpenPose: Real-Time Multi-Person Keypoint Detection Library for Body, Face, Hands, and Foot Estimation. Available online: https://github.com/CMU-Perceptual-Computing-Lab/openpose (accessed on 24 March 2020).
29. Wu, Y.; Kirillov, A.; Massa, F.; Lo, W.-Y.; Girshick, R. Detectron2. Available online: https://github.com/facebookresearch/detectron2 (accessed on 21 November 2020).
30. Cao, Z.; Simon, T.; Wei, S.-E.; Sheikh, Y. Realtime Multi-Person 2D Pose Estimation Using Part Affinity Fields. In Proceedings of the 2017 IEEE Conference on Computer Vision and Pattern Recognition (CVPR), Honolulu, HI, USA, 21–26 July 2017.
31. Viswakumar, A.; Rajagopalan, V.; Ray, T.; Parimi, C. Human Gait Analysis Using OpenPose. In Proceedings of the 2019 Fifth International Conference on Image Information Processing (ICIIP), Shimla, India, 15–17 November 2019; pp. 310–314. [CrossRef]
32. Pasinetti, S.; Muneeb Hassan, M.; Eberhardt, J.; Lancini, M.; Docchio, F.; Sansoni, G. Performance Analysis of the PMD Camboard Picoflexx Time-of-Flight Camera for Markerless Motion Capture Applications. *IEEE Trans. Instrum. Meas.* **2019**, *68*, 4456–4471. [CrossRef]
33. Albert, J.A.; Owolabi, V.; Gebel, A.; Brahms, C.M.; Granacher, U.; Arnrich, B. Evaluation of the Pose Tracking Performance of the Azure Kinect and Kinect v2 for Gait Analysis in Comparison with a Gold Standard: A Pilot Study. *Sensors* **2020**, *20*, 5104. [CrossRef] [PubMed]
34. Rosserial_Windows—ROS Wiki. Available online: http://wiki.ros.org/rosserial_windows (accessed on 14 March 2021).
35. OpenCV: Camera Calibration and 3D Reconstruction. Available online: https://docs.opencv.org/master/d9/d0c/group__calib3d.html (accessed on 14 March 2021).
36. Du, H.; Zhao, Y.; Han, J.; Wang, Z.; Song, G. Data Fusion of Multiple Kinect Sensors for a Rehabilitation System. In Proceedings of the Annual International Conference of the IEEE Engineering in Medicine and Biology Society, EMBS, Orlando, FL, USA, 16–20 August 2016; pp. 4869–4872. [CrossRef]

37. Jiang, Y.; Song, K.; Wang, J. Action Recognition Based on Fusion Skeleton of Two Kinect Sensors. In Proceedings of the 2020 International Conference on Culture-Oriented Science and Technology, ICCST 2020, Beijing, China, 28–31 October 2020; pp. 240–244. [CrossRef]
38. Human Pose Estimation by the OpenPose and Detectron 2—YouTube. Available online: https://www.youtube.com/watch?v=uwnrbqmns0Y (accessed on 24 March 2021).
39. Lavernia, C.; D'Apuzzo, M.; Rossi, M.D.; Lee, D. Accuracy of Knee Range of Motion Assessment After Total Knee Arthroplasty. *J. Arthroplast.* **2008**, *23*, 85–91. [CrossRef] [PubMed]

 *applied sciences*

Article

A BMI Based on Motor Imagery and Attention for Commanding a Lower-Limb Robotic Exoskeleton: A Case Study

Laura Ferrero *, Vicente Quiles, Mario Ortiz, Eduardo Iáñez and José M. Azorín

Brain-Machine Interface System Lab, Miguel Hernández University of Elche, 03202 Elche, Spain; vquiles@umh.es (V.Q.); mortiz@umh.es (M.O.); eianez@umh.es (E.I.); jm.azorin@umh.es (J.M.A.)
* Correspondence: lferrero@umh.es

Abstract: Lower-limb robotic exoskeletons are wearable devices that can be beneficial for people with lower-extremity motor impairment because they can be valuable in rehabilitation or assistance. These devices can be controlled mentally by means of brain–machine interfaces (BMI). The aim of the present study was the design of a BMI based on motor imagery (MI) to control the gait of a lower-limb exoskeleton. The evaluation is carried out with able-bodied subjects as a preliminary study since potential users are people with motor limitations. The proposed control works as a state machine, i.e., the decoding algorithm is different to start (standing still) and to stop (walking). The BMI combines two different paradigms for reducing the false triggering rate (when the BMI identifies irrelevant brain tasks as MI), one based on motor imagery and another one based on the attention to the gait of the user. Research was divided into two parts. First, during the training phase, results showed an average accuracy of $68.44 \pm 8.46\%$ for the MI paradigm and $65.45 \pm 5.53\%$ for the attention paradigm. Then, during the test phase, the exoskeleton was controlled by the BMI and the average performance was $64.50 \pm 10.66\%$, with very few false positives. Participants completed various sessions and there was a significant improvement over time. These results indicate that, after several sessions, the developed system may be employed for controlling a lower-limb exoskeleton, which could benefit people with motor impairment as an assistance device and/or as a therapeutic approach with very limited false activations.

Keywords: brain–machine interfaces; EEG; exoskeleton; motor imagery

Citation: Ferrero, L.; Quiles, V.; Ortiz, M.; Iáñez, E.; Azorín, J.M. A BMI Based on Motor Imagery and Attention for Commanding a Lower-Limb Robotic Exoskeleton: A Case Study. *Appl. Sci.* **2021**, *11*, 4106. https://doi.org/10.3390/app11094106

Academic Editor: Carlos A. Jara

Received: 28 February 2021
Accepted: 29 April 2021
Published: 30 April 2021

Publisher's Note: MDPI stays neutral with regard to jurisdictional claims in published maps and institutional affiliations.

1. Introduction

Robotic exoskeletons are wearable devices that can enhance physical performance and provide movement assistance. In the case of lower-limb robotic exoskeletons, they can be beneficial for people with motor impairment in the lower extremities as they can assist the gait and facilitate rehabilitation [1]. The combination of lower-limb robotic exoskeletons with brain–machine interfaces (BMI), which are systems that decode neural activity to drive output devices, offers a new method to provide motor support. Thus, patients could walk while being assisted by an exoskeleton that is controlled by their brain activity.

In the literature, there are different BMI control paradigms for lower-limb exoskeletons based on brain changes. The most common ones are steady-state visually evoked potentials [2], which are based on visual stimuli; motion-related cortical potentials [3–6], which are produced between 1500 and 500 ms before the execution of the movement, and and event-related desynchronization/synchronization (ERD/ERS), which is considered to indicate the activation and posterior recovery of the motor cortex during preparation and completion of a movement [7–9]. BMI based on ERD/ERS are usually employed to detect motion intention [3,6,10]. Similar ERD/ERS patterns are produced during motor imagery (MI), which consists of the imagination of a movement [11–13]. When performing MI, in contrast to external stimuli, brain changes are induced voluntarily and internally by the subject. BMI based on MI have the objective of identifying different MI tasks or differentiating between MI and an idle state [5,14–16]. The work of [16] combined MI with eye blinks as a control criterion.

The main limitation of MI is that patients have to maintain it for long periods in order to force the external device to perform any action. However, contrary to instantaneous brain changes, such as MRCP or motion intention, continuous cognitive involvement of a patient during the assisted motion can induce mechanisms of neuroplasticity. Neuroplasticty is the ability of the brain to reorganize its structure and promote rehabilitation [17]. The performance of maintained brain tasks can be challenging as it requires high focus from the user during the whole experiment and any external influence could easily disturb it. Previous studies have tried to evaluate the level of attention of a subject during the control of the external device [18] and some of them have considered it as a control paradigm for a lower-limb BMI [15]. BMI systems need a training phase in which the model is calibrated for each subject and then it is tested with with new data. In [5,14–16], during the training phase, participants alternated periods of MI with idle state and the output device was only moving during MI. Nevertheless, since BMI focus on sensorimotor rhythms, it is difficult to ensure that it is not considering the actual motion instead of motor imagery.

In our previous work [19], we designed a lower-limb MI BMI to control a treadmill and it was tested with able-bodied subjects. The BMI combined the paradigm of MI with another one that measured the level of attention that users had during MI tasks. In the test phase, i.e., when the output device was commanded by the BMI, the treadmill was only activated when the attention measured was higher than a certain threshold, reducing the number of false triggers. In order to ensure that motion artifacts did not affect the BMI classifier model, the training phase consisted of two types of trials: full standing and full motion trials. The mental tasks to perform were the same for both types, alternating periods of MI with idle state. Both types of trials allowed the creation of two different classifier models to be applied depending on the status of the subject: gait and stand.

In this study, the BMI designed in [19] was adapted for the control of the gait of a lower-limb exoskeleton and it was evaluated with able-bodied subjects. The combination of this BMI with a lower-limb exoskeleton is a promising and intuitive assistive approach for people with motor impairment. In addition, it could potentially benefit people with cortical damage (e.g., after a stroke) as a therapeutic approach for the recovery of lost motor function. Participants were trained over 2–5 days to assess the effect of practice on the performance. Each day's session was divided into two parts: the training and test phases. During training, subjects performed trials in which the exoskeleton was walking the entire time and trials in which it was standing. In the test phase, the exoskeleton provided real-time feedback in a closed-loop control scenario. This is a previous step in the development of a BMI that will reinforce rehabilitation and/or assist the gait for patients with neurological damage.

2. Materials and Methods

2.1. Participants

Two subjects participated in the study (mean age 23.5 ± 3.5). They did not report any known disease and had no movement impairment. They did not have any previous experience with BMI. They were informed about the experiments and signed an informed consent form in accordance with the Declaration of Helsinki. All procedures were approved by the Responsible Research Office of Miguel Hernández University of Elche.

2.2. Equipment

Brain activity was recorded with electroencephalography (EEG). A 32-electrode system actiCap (Brain Products GmbH, Germany) was employed to record EEG signals. The 27 channels selected for acquisition were: F3, FZ, FC1, FCZ, C1, CZ, CP1, CPZ, FC5, FC3, C5, C3, CP5, CP3, P3, PZ, F4, FC2, FC4, FC6, C2, C4, CP2, CP4, C6, CP6, P4. They were placed following the 10-10 international system on an actiCAP (Brain Products GmbH, Germany). Four electrodes were located next to the eyes to record electrooculography

(EOG) and ground and reference electrodes were located on the right and left ear lobes, respectively. Each channel signal was amplified with BrainVision BrainAmp amplifier (Brain Products GmbH, Germany). Finally, signals were transmitted wirelessly to the BrainVision recorder software (Brain Products GmbH, Germany).

H3 exoskeleton (Technaid, Madrid, Spain) was employed to assist the movement and participants used crutches as support. Control start/stop gait commands were sent via Bluetooth. The experimental setup can be seen in Figure 1.

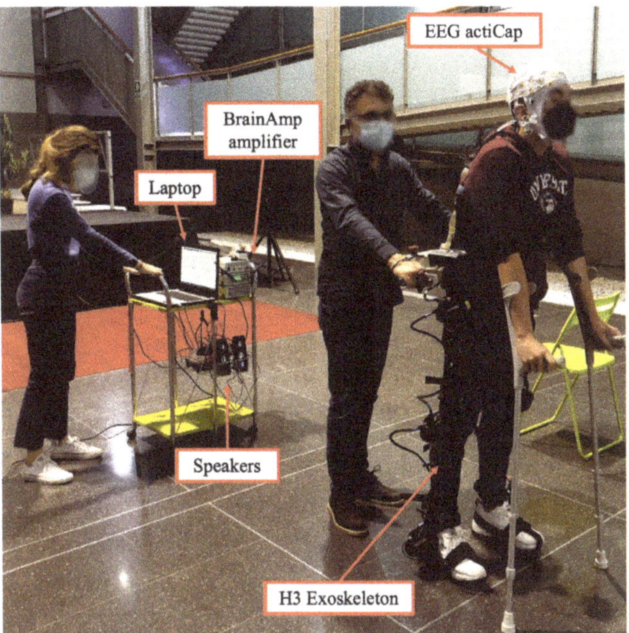

Figure 1. Experimental setup.

2.3. Experimental Design

Each participant completed several sessions and each session was divided into two parts. The first part consisted of the training phase, in which the exoskeleton was in opened-loop control. Thus, it was remotely controlled by the laptop with predefined commands based on the mental tasks to be registered and not by the output of the BMI classifier. Afterwards, the second part of the session allowed assessment of the BMI performance during closed-loop control of the exoskeleton. Commands issued by the BMI were sent to the exoskeleton in real time based on the decoding of the brain activity obtained as output of the BMI classifier, receiving the subjects' real-time feedback on their performance.

2.3.1. Training Phase

In the first part of each session, subjects performed 20 trials. Each trial consisted of a sequence of three mental tasks: MI of the gait, idle state and regressive count. For idle state, participants were asked to be as relaxed as possible. The regressive count was randomly changed every trial and consisted of a number between 300 and 1000 and a subtrahend between 1 and 9. For example, if they were given the count 500-4, they had to compute the series of subtractions of 496, 492, 488... until they had to perform the following task. This task aims to focus the subject on a demanding mental task very different to MI in order to assess a low level of attention to gait. The protocol can be seen in Figure 2a. There was a voice message that indicated the beginning of each task: 'Relax', 'Imagine', '500-5'. The message for the regressive count indicated a different mathematical operation each time. In order to avoid evoked potentials, the 4 s period after auditory cues was not considered for further analysis.

During the session, subjects used crutches to maintain stability. In addition, a member of the research staff softly held the exoskeleton to prevent any possible loss of balance or fall. Ten of the training trials were performed in a full no-motion status and the other ten in a full motion status assisted by the exoskeleton. These trials were employed to train two different BMI classifiers: StandClassifiers (with non-motion trials) and GaitClassifiers (with full motion trials).

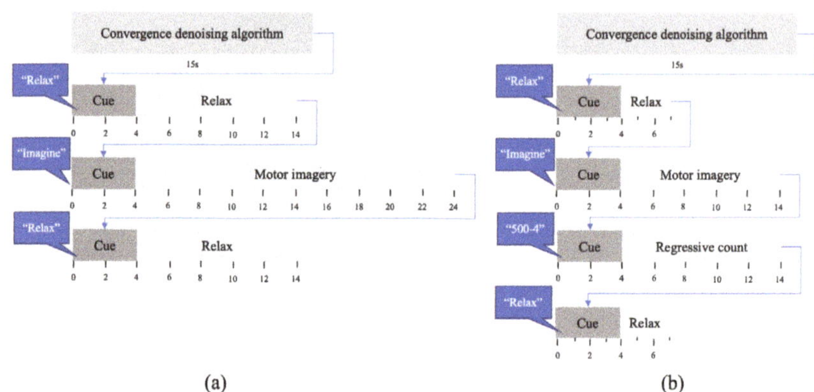

Figure 2. (**a**) The protocol of opened-loop trials and (**b**) the protocol of close-loop trials.

2.3.2. Test Phase

In the second part of each session, the BMI was tested in closed-loop control with the two groups of classifiers obtained with the data of the training phase (StandClassifiers, GaitClassifiers). Subjects performed five trials, whose protocol can be seen in Figure 2b. The transition between tasks was indicated with voice messages for 'Relax' and 'Imagine' tasks. Notice that no 'Regression count' task was considered, as attention level to gait was computed based on the information from training, but there was no need to implement a low-level gait attention task in the testing trials.

2.4. Brain Machine Interface

The presented BMI had the following steps: data acquisition, pre-processing, feature extraction, classification, exoskeleton control and evaluation.

As indicated before, this BMI was based on two paradigms: MI and attention. The first one was based on the distinction between MI of the gait and an idle state, so only data associated with these brain tasks were considered to train the classifiers (relax and motor imagery). With regard to the attention paradigm, it measured the level of attention to gait. Therefore, it had the objective of differentiating between the attention of the subject during MI and the attention during irrelevant tasks. For this paradigm, all brain tasks from training trials were contemplated (relax, motor imagery and regressive count). While the attention to the gait was assumed to be high during MI tasks, it was assumed to be low during regressive count and idle state. The schema of the BMI can be seen in Figure 3.

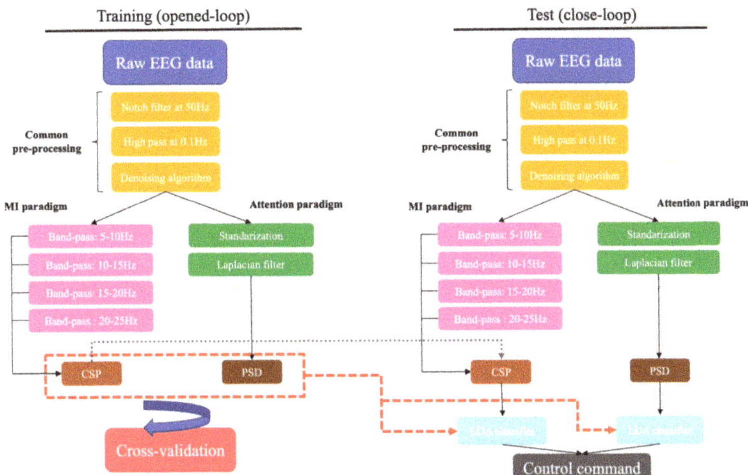

Figure 3. Brain–machine interface (BMI) scheme. During training, the exoskeleton was in opened-loop control, and for testing, it was in closed-loop control. The BMI used two different paradigms: one based on motor imagery of gait and another one based on the user's level of attention to gait. Both paradigms shared some steps of pre-processing but there were additional different steps for each one. Then, two different feature extraction methods were employed. Trials from the training phase were used to train the BMI classifiers for testing.

2.4.1. Data Acquisition

EEG signals were recorded at a sampling frequency of 200 Hz. Then, epochs of 1 s with 0.5 s of shifting were extracted and processed.

2.4.2. Pre-Processing

The pre-processing stage started with two frequency filters: a notch filter at 50 Hz to remove the contribution of the power line and a high-pass filter at 0.1 Hz. In order to reduce motion artifacts, electrode wires were fixed with clamps and a medical mesh. The movement of jaw muscles can generate signal artifacts, so subjects were asked to not swallow or chew while they were performing MI, regressive count or were in a idle state.

The H_∞ denoising algorithm was applied to mitigate the presence of eye artifacts and signal drifts [5]. This algorithm estimates the contribution of the EOG and a constant parameter to the EEG signal and removes it. Afterwards, there were two different pre-processing lines, one for each paradigm.

For the MI paradigm, a filter bank comprising multiple band-pass filters was applied to the data after the H_∞ denoising algorithm. Four band-pass filters were employed to obtain data associated with alpha and beta rhythms.

Regarding the attention paradigm, EEG signals from each channel were first standardized following the process presented in [20]. For each channel, the maximum visual threshold was computed as the mean of the 6 highest values of the signal. This value was iteratively updated for each epoch and it was used to standardize the data as

$$SV(t)_{ch} = \frac{V(t)_{ch}}{\frac{1}{Ch}\sum_{j=1}^{Ch} MVThreshold_j}. \tag{1}$$

The signal of each chanel, $V(t)_{ch}$, was normalized taking into consideration the maximum visual threshold ($MVThreshold$) of all the EEG channels. Subsequently, the surface Laplacian filter was used to reduce spatial noise and enhance the local activity of each electrode [21].

2.4.3. Feature Extraction

The following step of the BMI has the objective of computing the characteristics of the EEG during each brain task that could be discriminating.

For the MI paradigm, common spatial patterns (CSP) [22] are computed for each frequency band. CSP estimate a spatial transformation that maximizes the discriminability between two brain patterns. If X is the EEG that has $N * T$ dimensions, which are the number of channels and number of samples, respectively, the CSP algorithm estimates a matrix of spatial filters W that discriminates between two classes: (X_1) and (X_2). Firstly, the normalized covariance matrices are computed for each class as in

$$C_1 = \frac{X_1 X_1^T}{trace(X_1 X_1^T)}, C_2 = \frac{X_2 X_2^T}{trace(X_2 X_2^T)}. \tag{2}$$

These matrices are computed for each trial and $\overline{C_1}$ and $\overline{C_2}$ are calculated by averaging over all trials of the same class. The averaged covariance matrices are combined to result in the composite spatial covariance matrix that can be factorized as

$$C = \overline{C_1} + \overline{C_2} = U_0 \Sigma U_0^T. \tag{3}$$

U_0 is a matrix of eigenvectors and Σ is the diagonal matrix of eigenvalues. The averaged covariance matrices are transformed as

$$P = \Sigma^{1/2} U_0^T, \tag{4}$$

$$S_1 = P\overline{C_1}P^T, S_2 = P\overline{C_2}P^T. \tag{5}$$

S_1 and S_2 have common eigenvectors, and the sum of both matrices of eigenvalues is the identity matrix.

$$S_1 = U\Sigma_1 U^T, S_1 = U\Sigma_2 U^T and \Sigma_1 + \Sigma_2 = I \tag{6}$$

The projection matrix is obtained as

$$W = U^T P. \tag{7}$$

Z is the projection of the original EEG signal S into another space. Columns of W^{-1} are the spatial patterns.

$$Z = WX \tag{8}$$

Although Z has $N * T$ dimensions, the first and last rows are the components that can be better discriminated in terms of their variance. Therefore, for feature extraction, only the m first and last components of Z are considered. Z_p is the subset of Z and the variances of each component are computed and normalized with the logarithm as

$$f_p = log \frac{var(Z_p)}{\sum_{i=1}^{2m} Z_p}. \tag{9}$$

f_p is the vector of features and has $(fbands * 2 * m) * T$ dimension. m was set to 4, and in the pre-processing phase, 4 band-pass filters were employed so the dimension is $32 * T$.

For the attention paradigm, power spectral estimation by Maximum Entropy Method (MEM) was used to obtain features associated with each task. The signal of each electrode was estimated as an autoregressive model in which the known autocorrelation coefficients were calculated and the unknown coefficients were estimated by maximizing the spectral entropy [23]. Afterwards, the autocorrelation cofficients were used to compute the power spectrum that was compatible with the fragment of the signal analyzed, but it was also evasive regarding unseen data. Afterwards, only the power of the frequencies in the gamma band was considered [15].

2.4.4. Classification

Training trials of each session were evaluated using leave-one-out cross-validation. Each trial was once used as a test and the remaining trials conformed to the training group. This process was performed independently for trials in which subjects were standing (10 trials) and trials in which they were in motion (10 trials). Linear Discriminant Analysis (LDA) [24] classifiers were created depending on the subject status—full standing trials (StandClassifiers) and full motion trials(GaitClassifiers)—each one with two different models based on the decoding paradigm: MI and attention paradigms. As stated above, whereas LDA classifiers of the MI paradigm were only trained with data from MI and idle state, LDA classifiers of the attention paradigm were trained with data from all brain tasks (idle, regressive count, MI).

Concerning the test phase, the developed BMI was designed as a state machine system in which a group of classifiers was chosen based on the status of the exoskeleton. This way, if the subject is in a standing position, the MI and attention classifiers of the full standing trials (StandClassifiers) are used to decide if the exoskeleton keeps standing or starts moving, but if the subject is moving, the MI and attention classifiers obtained by the full motion trials (GaitClassifiers) are used to continue walking or to stop. Predictions from both paradigms were combined to decode control commands. Its design can be seen in Figure 4. In summary, in each test trial, subjects started standing with the exoskeleton and StandClassifiers were employed. The system could decode stop or walk commands based on the prediction of their MI and attention classifiers. When a walk command was sent to the exoskeleton, it started the gait and the system was changed to Gait state. Consequently, GaitClassifiers were employed afterwards. Again, the system could decode stop or walk commands, but when a stop command was issued, the exoskeleton stopped the gait and the system changed to Stand state again.

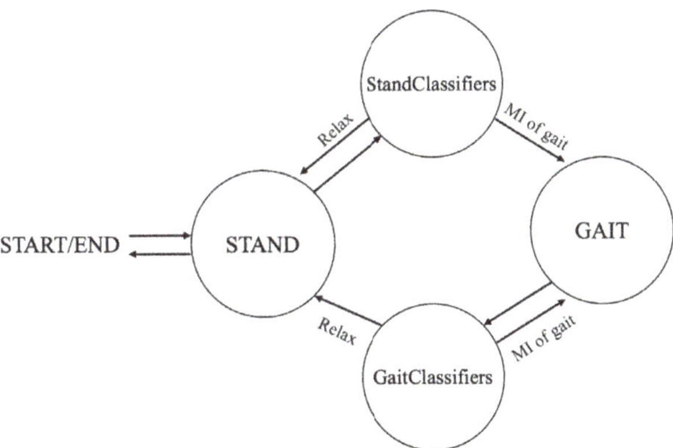

Figure 4. State machine design of the brain–machine interface (BMI). There are two states, gait and stand, that depend on the exoskeleton status. Each state is associated with two different classifiers, one for each paradigm, that will be used to give decode control commands.

2.4.5. Exoskeleton Control

In the test phase, the exoskeleton was controlled by BMI decoded commands. MI classifiers could predict two classes, 0 for idle state and 1 for MI, and attention classifiers could predict a 0 for low attention to gait and a 1 for high attention. These predictions were averaged every 10 s, which resulted in MI and attention indices that ranged from 0 to 1. Control commands were selected based on the following rules:

- During 5 s, new commands cannot be issued.
- If subject was standing:

- If the MI index was higher than or equal than 0.7 or the MI index was higher than or equal than 0.6 and the attention index was higher than or equal to 0.4, a move command was issued and the exoskeleton started the gait.
- Otherwise, the exoskeleton kept standing.

- If the subject was walking:
  - If the MI index was lower than or equal to 0.4, a stop command was issued and the exoskeleton stopped the gait.
  - Otherwise, it kept walking.

2.5. Evaluation

The accuracy of training trials was defined as the percentage of correctly classified epochs during each brain task. This metric was computed separately for trials in which participants were moving and trials in which they were static. Furthermore, the performance of closed-loop trials was assessed with the following indices:

- %MI and %Att: percentage of epochs of data correctly classified for each paradigm.
- %Commands: percentage of epochs of data with correct control commands.
- Accuracy commands: percentage of correct commands issued.
- True positive ratio (TPR): percentage of MI periods in which a walking event is executed. There is only an event of MI per trial, so this value can only be 0 or 100% per trial.
- False positives (FP) and false positives per minute (FP/min): moving commands issued during rest periods.

Transition events were not considered for the computation of evaluation metrics.

3. Results

During training, participants wore the exoskeleton in an opened-loop control. Each subject completed several sessions, and on each of them, they completed 20 trials: 10 trials standing still and 10 trials walking. Results from subjects S1 and S2 are shown in Tables 1 and 2, respectively. It must be noted that they did not have the same amount of practice since they participated in a different number of sessions. Two different BMI paradigms were carried out. For the MI paradigm, S1 reached an average accuracy of $72.77 \pm 6.61\%$ with a difference of around 6% between the two conditions, standing and walking. In the last session, S2 achieved an average accuracy of 64.11 ± 9.98 with a difference of 20% between the two approaches. With respect to the attention paradigm, S1 obtained an accuracy of 65.06 ± 6.44 with a difference of 8%, and S2 achieved 65.83 ± 4.43 and a 10% difference. The average accuracy of the MI and attention paradigm was $68.44 \pm 8.46\%$ and $65.45 \pm 5.53\%$, respectively.

Figures 5 and 6 show the spatial patterns of S1 and S2 in their last session. Moreover, in order to provide a comparison under the same conditions, Figure 7 shows the spatial patterns of S3 in the second session. The spatial patterns estimated during trials without movement show that for S1 and S2, electrode FCz seems to have a relevant role in the discrimination of idle state. During MI events, the most significant electrodes for both subjects are peripheral as FC5. However, results from S2 show that in the 5–10 Hz band, C2 and CPz are relevant to the MI of gait. Regarding trials in which participants are walking, the distribution of relevant areas seems scattered for idle state and for MI; peripheral electrodes are also highlighted.

When comparing the spatial patterns of S2 in two different sessions, the main similarities can be found in the stand trials. CPz and Cz are highlighted for the relax class and electrode FC5 seems to be significant for the MI class.

3.1. Training Phase

Table 1. Results from training, subject S1. Trials with opened-loop control of the exoskeleton.

		Session 1	Session 2
Stand	%MI	59.29 ± 10.51	69.64 ± 7.62
	%Att	57.38 ± 9.27	60.83 ± 7.58
Gait	%MI	58.93 ± 11.60	75.89 ± 5.41
	%Att	65.83 ± 8.57	69.29 ± 5.04

Table 2. Results from training, subject S2. Trials with opened-loop control of the exoskeleton.

		Session 1	Session 2	Session 3	Session 4	Session 5
Stand	%MI	53.32 ± 8.59	69.64 ± 8.70	65.54 ± 5.12	64.2 ± 11.45	74.11 ± 6.14
	%Att	63.95 ± 4.44	62.57 ± 8.77	58.21 ± 8.76	59.52 ± 8.44	60.83 ± 5.48
Gait	%MI	50.26 ± 7.54	54.17 ± 8.75	62.50 ± 8.91	59.82 ± 10.11	54.11 ± 12.71
	%Att	61.05 ± 5.13	63.36 ± 2.84	61.55 ± 7.21	65.71 ± 6.82	70.83 ± 3.04

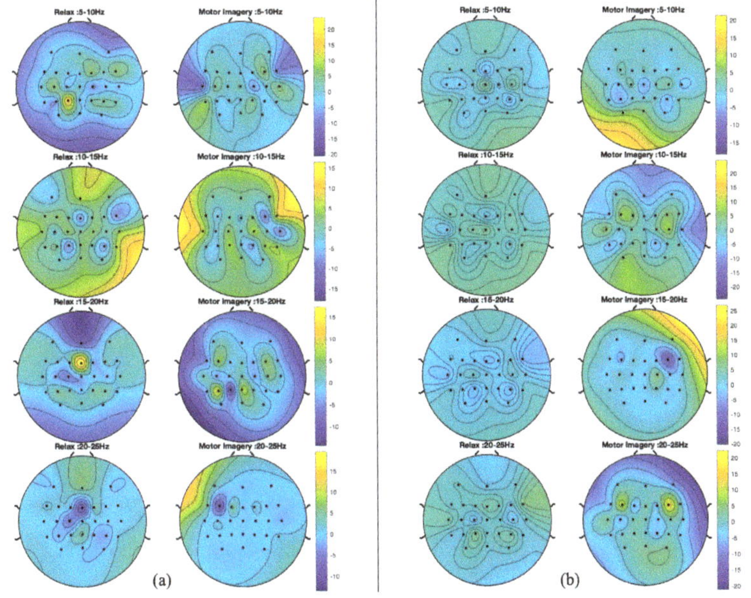

Figure 5. Spatial patterns for the session of S1 that best discriminate between motor imagery (MI) and idle state. (**a**) The spatial patterns from trials in which participant was standing still and (**b**) the spatial patterns from trials in which they were walking with the exoskeleton.

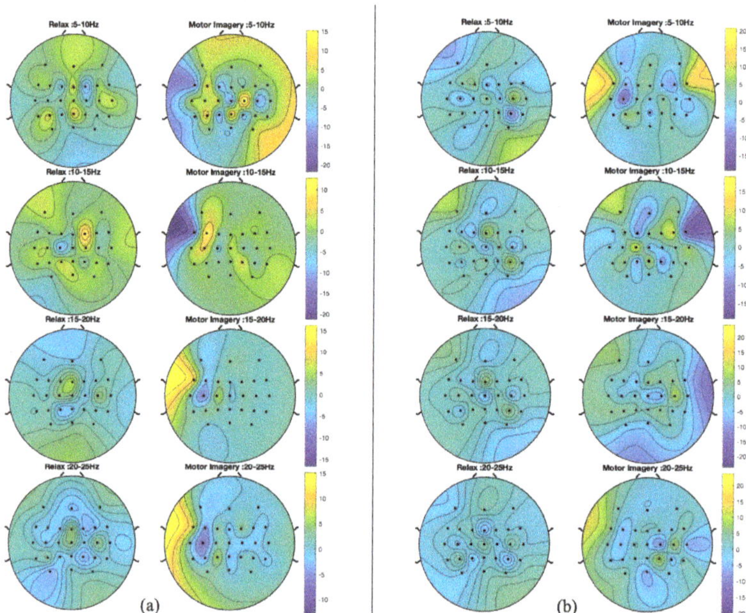

Figure 6. Spatial patterns for the session of S2 that best discriminate between motor imagery (MI) and idle state. (**a**) The spatial patterns from trials in which participant was standing still and (**b**) the spatial patterns from trials in which they were walking with the exoskeleton.

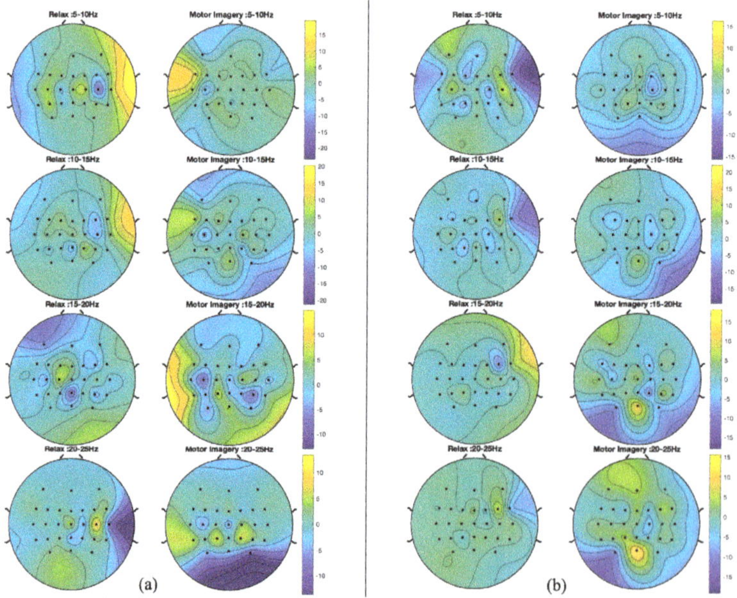

Figure 7. Spatial patterns for the fifth session of S2 that best discriminate between motor imagery (MI) and idle state. (**a**) The spatial patterns from trials in which participant was standing still and (**b**) the spatial patterns from trials in which they were walking with the exoskeleton.

3.2. Test Phase

The exoskeleton was controlled by the BMI decoded commands and the BMI classifiers were trained with training trials. Tables 3–5 summarize the results from closed-loop trials. TPR is 100% in the majority of trials, which means that the exoskeleton was activated at least once during the MI event. The number of false positive activations during idle state ranged from 0 to 2. Regarding %Commands, it improved by 13% from the first to the last session of S2, although their performance in each session was not always superior to the previous one. In the last session, the average %Commands for both subjects was 64.50 ± 10.66%.

Table 3. Test results, subject S1. Trials in close-loop control.

		Trial 1	Trial 2	Trial 3	Trial 4	Trial 5	Avg.
	%MI	64.13	50.00	48.91	61.96	56.52	56.30
	%Att	51.09	58.70	53.26	47.83	55.43	53.26
	%Commands	63.00	78.00	62.00	60.00	83.00	69.20
Session 1	Acc. commands	50.00	50.00	50.00	0.00	50.00	40.00
	TPR	100.00	100.00	100.00	0.00	100.00	80.00
	FP	1.00	0.00	1.00	0.00	0.00	0.40
	FP/min	2.31	0.00	2.31	0.00	0.00	0.92
	%MI	61.96	52.17	50.00	60.87	57.61	56.52
	%Att	64.13	46.74	61.96	60.87	57.61	58.26
	%Commands	60.00	57.00	63.00	57.00	69.00	61.20
Session 2	Acc. commands	75.00	75.00	75.00	50.00	66.67	68.33
	TPR	100.00	100.00	100.00	100.00	100.00	100.00
	FP	1.00	1.00	1.00	1.00	1.00	1.00
	FP/min	2.31	2.31	2.31	2.31	2.31	2.31

Table 4. Test results, first two sessions of subject S2. Trials in close-loop control.

		Trial 1	Trial 2	Trial 3	Trial 4	Trial 5	Avg.
	%MI	50	44.57	43.48	42.39	43.48	44.78
	%Att	47.83	56.52	53.26	53.26	50	52.17
	%Commands	59.00	54.00	53.00	53.00	53.00	54.40
Session 1	Acc. commands	0.00	0.00	0.00	0.00	0.00	0.00
	TPR	100.00	100.00	100.00	100.00	100.00	100.00
	FP	1.00	1.00	1.00	1.00	1.00	1.00
	FP/min	2.31	2.31	2.31	2.31	2.31	2.31
	%MI	46.74	60.87	48.91	54.35	52.17	52.61
	%Att	56.52	59.78	43.48	66.3	64.13	58.04
	%Commands	59.00	53.00	37.00	76.00	68.00	58.60
Session 2	Acc. commands	0.00	100.00	0.00	66.67	100	53.33
	TPR	100.00	100.00	0.00	100.00	100.00	80.00
	FP	1.00	0.00	1.00	1.00	0.00	0.60
	FP/min	2.31	0.00	2.31	2.31	0.00	1.38

Table 5. Test results, last three sessions of subject S2. Trials in close-loop control.

		Trial 1	Trial 2	Trial 3	Trial 4	Trial 5	Avg.
Session 3	%MI	63.04	46.74	52.17	48.91	46.74	51.52
	%Att	45.65	44.57	55.43	52.17	50	49.56
	%Commands	67.00	81.00	75.00	63.00	63.00	69.80
	Acc. commands	60.00	60.00	75.00	40.00	50.00	57.00
	TPR	100.00	100.00	100.00	100.00	100.00	100.00
	FP	2.00	1.00	1.00	2.00	2.00	1.60
	FP/min	4.62	2.31	2.31	4.62	4.62	3.69
Session 4	%MI	53.26	64.13	45.65	58.7	58.7	56.09
	%Att	58.7	56.52	75	55.43	71.74	63.48
	%Commands	57.00	64.00	57.00	78.00	65.00	64.20
	Acc. commands	0.00	100.00	0.00	75.00	50.00	45.00
	TPR	100.00	100.00	100.00	100.00	100.00	100.00
	FP	1.00	0.00	1.00	1.00	1.00	0.80
	FP/min	2.31	0.00	2.31	2.31	2.31	1.85
Session 5	%MI	59.78	56.52	59.78	59.78	61.96	59.56
	%Att	70.65	70.65	59.78	67.39	56.52	65.00
	%Commands	56.00	73.00	52.00	88.00	70.00	67.80
	Acc. commands	40.00	66.67	100.00	100.00	33.33	68.00
	TPR	100.00	100.00	100.00	100.00	100.00	100.00
	FP	2.00	1.00	0.00	0.00	2.00	1.00
	FP/min	4.62	2.31	0.00	0.00	4.62	2.31

4. Discussion

Contrary to the findings of our previous work on a BMI-controlled treadmill [19], we found significant differences between opened-loop trials in which subjects were standing and when they were walking. It is important to note that walking assisted by an exoskeleton is a more complex task than walking on a treadmill, so subjects must be concentrated. Consequently, it is more difficult for them to perform other brain tasks such as MI or regressive count. In addition, when comparing the results from closed-loop trials, the average percentage of epochs with correct commands was 64.5% with the exoskeleton and 75.6% with the treadmill. A possible explanation for this contrast could be also related to the complexity of the movement with the exoskeleton.

On the other hand, the attention paradigm showed worse performance than the MI paradigm in opened-loop trials, which is consistent with the findings of our previous work [19]. However, in line with our previous work with an exoskeleton [15], this difference is not as evident in closed-loop trials. Therefore, future BMI designs could rely more on the attention paradigm for the activation of the exoskeleton.

While results from the MI paradigm showed an increasing trend throughout sessions, this pattern is not as evident for the attention paradigm. Our results for the MI paradigm are in consonance with the conclusions from [25]. Performing MI is not an intuitive activity for novel participants and practice could promote the modulation and enhance brain activity patterns. Nevertheless, with regard to the attention of the user, the performance does not seem to improve with practice. The attention is something that people train on daily basis, so this could explain why a few sessions cannot further improve it.

There are not many investigations in the literature that developed BMI based on lower-limb MI without other external stimuli [2] and they are usually based on motion intention [3,6,10]. In addition, the works of [4,26] employed upper-limb MI to control a lower-limb exoskeleton. Reference [26] showed a percentage of correct commands issued every 4.5 s of 66% and [4] of 80.16% but the BMI was only employed to start the gait and not to stop it. These values can be compared with the %Commands of the present paper. Although superior results are achieved with upper-limb MI, this paradigm cannot be applied to promote neuroplasticity.

In [16], a BMI was presented that employed a combination of MI with eye blinking as a control paradigm, and an accuracy of 86.7% was reported. However, although control mechanisms that employ eye movements have proven to be precise, they lack application from the rehabilitation point of view. In addition, the work of [14] presented a BMI that only controlled the start and maintenance of the gait of a lower-limb exoskeleton and they obtained an average accuracy of 74.4%. In our previous research [15] that also combined the MI and attention paradigms to control an exoskeleton, the percentage of epochs with correct commands issued was 56.77%. Slightly superior results were achieved with the current BMI algorithm.

5. Conclusions

The current research presents a BMI system based on MI and attention paradigms that has been tested to control a lower-limb exoskeleton. Participants performed 2–5 sessions to assess the effect of practice on the performance. Each session was divided into two parts: the training and test phases. First, participants completed trials in which they had to perform certain brain tasks and the exoskeleton was controlled remotely by the laptop with predefined commands. During half of the trials, the exoskeleton was walking, and during the other half, it was completely static. Therefore, contrary to previous works, brain tasks to discriminate happened under the same conditions. Moreover, this setup can reduce the effect of artifacts on the predictions. The average performance in the last session was $68.44 \pm 8.46\%$ for the MI paradigm and $65.45 \pm 5.53\%$ for the attention paradigm. The second part of the each session consisted of closed-loop controlled trials in which the exoskeleton was commanded by the predictions of the BMI. The BMI worked as a state machine that used different classifiers depending on whether the exoskeleton was static or moving. Training trials were used to train the classifiers corresponding to each state of the state machine. The BMI took a decision every 0.5 s and the average percentage of correct commands chosen was $64.50 \pm 10.66\%$ for the last session of both subjects.

Participants did not have any motor impairment, but since the main of objective of the system is to promote neurorehabilitation and neuroplasticity, future research will focus on people with motor disabilities.

Author Contributions: Conceptualization, L.F. and M.O.; methodology, L.F., V.Q. and M.O.; software, L.F. and V.Q.; validation, M.O. and E.I.; formal analysis, L.F. and V.Q.; investigation, L.F. and V.Q.; resources, M.O., E.I. and J.M.A.; data curation, E.I.; writing—original draft preparation, L.F.; writing—review and editing, L.F. and M.O.; visualization, L.F.; supervision, M.O., E.I. and J.M.A.; project administration, J.M.A.; funding acquisition, J.M.A. All authors have read and agreed to the published version of the manuscript.

Funding: This research was funded by the Spanish Ministry of Science and Innovation, the Spanish State Agency of Research, and the European Union through the European Regional Development Fund in the framework of the project Walk—Controlling lower-limb exoskeletons by means of brain–machine interfaces to assist people with walking disabilities (RTI2018-096677-B-I00); and by the Consellería de Innovación, Universidades, Ciencia y Sociedad Digital (Generalitat Valenciana), and the European Social Fund in the framework of the project Desarrollo de nuevas interfaces cerebro-máquina para la rehabilitación de miembro inferior (GV/2019/009).

Institutional Review Board Statement: The study was conducted according to the guidelines of the Declaration of Helsinki, and approved by the Institutional Review Board of Miguel Hernandez University of Elche (DIS.JAP.03.18 and 22/01/2019).

Informed Consent Statement: Informed consent was obtained from all subjects involved in the study.

Data Availability Statement: Not applicable.

Conflicts of Interest: The authors declare no conflict of interest.

References

1. Bogue, R. Robotic exoskeletons: A review of recent progress. *Ind. Robot.* **2015**, *42*, 5–10. [CrossRef]
2. Kwak, N.S.; Müller, K.R.; Lee, S.W. A convolutional neural network for steady state visual evoked potential classification under ambulatory environment. *PLoS ONE* **2017**, *12*, e0172578. [CrossRef] [PubMed]
3. Zhang, Y.; Prasad, S.; Kilicarslan, A.; Contreras-Vidal, J.L. Multiple kernel based region importance learning for neural classification of gait states from EEG signals. *Front. Neurosci.* **2017**, *11*, 170. [CrossRef]
4. Liu, D.; Chen, W.; Pei, Z.; Wang, J. A brain-controlled lower-limb exoskeleton for human gait training. *Rev. Sci. Instrum.* **2017**, *88*, 104302. [CrossRef] [PubMed]
5. Kilicarslan, A.; Grossman, R.G.; Contreras-Vidal, J.L. A robust adaptive denoising framework for real-time artifact removal in scalp EEG measurements. *J. Neural Eng.* **2016**, *13*, 026013. [CrossRef] [PubMed]
6. Kilicarslan, A.; Prasad, S.; Grossman, R.G.; Contreras-Vidal, J.L. High accuracy decoding of user intentions using EEG to control a lower-body exoskeleton. In Proceedings of the 2013 35th Annual International Conference of the IEEE Engineering in Medicine and Biology Society (EMBC), Osaka, Japan, 3–7 July 2013; pp. 5606–5609. [CrossRef]
7. Pfurtscheller, G.; Neuper, C.; Flotzinger, D.; Pregenzer, M. EEG-based discrimination between imagination of right and left hand movement. *Electroencephalogr. Clin. Neurophysiol.* **1997**, *103*, 642–651. [CrossRef]
8. Pfurtscheller, G.; Lopes Da Silva, F.H. Event-related EEG/MEG synchronization and desynchronization: Basic principles. *Clin. Neurophysiol.* **1999**, *110*, 1842–1857. [CrossRef]
9. Seeland, A.; Manca, L.; Kirchner, F.; Kirchner, E.A. Spatio-temporal comparison between ERD/ERS and MRCP-based movement prediction. In Proceedings of the BIOSIGNALS 2015—8th International Conference on Bio-Inspired Systems and Signal Processing, Lisbon, Portugal, 12–15 January 2015; pp. 219–226. [CrossRef]
10. Rajasekaran, V.; López-Larraz, E.; Trincado-Alonso, F.; Aranda, J.; Montesano, L.; Del-Ama, A.J.; Pons, J.L. Volition-adaptive control for gait training using wearable exoskeleton: Preliminary tests with incomplete spinal cord injury individuals. *J. Neuroeng. Rehabil.* **2018**, *15*, 4. [CrossRef]
11. Stippich, C.; Ochmann, H.; Sartor, K. Somatotopic mapping of the human primary sensorimotor cortex during motor imagery and motor execution by functional magnetic resonance imaging. *Neurosci. Lett.* **2002**, *331*, 50–54. [CrossRef]
12. Bakker, M.; de Lange, F.P.; Stevens, J.A.; Toni, I.; Bloem, B.R. Motor imagery of gait: A quantitative approach. *Exp. Brain Res.* **2007**, *179*, 497–504. [CrossRef]
13. Batula, A.M.; Mark, J.A.; Kim, Y.E.; Ayaz, H. Comparison of Brain Activation during Motor Imagery and Motor Movement Using fNIRS. *Comput. Intell. Neurosci.* **2017**, *2017*, 5491296. [CrossRef]
14. Rodríguez-Ugarte, M.; Iáñez, E.; Ortiz, M.; Azorín, J.M. Improving Real-Time Lower Limb Motor Imagery Detection Using tDCS and an Exoskeleton. *Front. Neurosci.* **2018**, *12*, 757. [CrossRef] [PubMed]
15. Ortiz, M.; Iáñez, E.; Gaxiola, J.; Kilicarslan, A.; Azorín, J.M.; Member, S. Assessment of motor imagery in gamma band using a lower limb exoskeleton. In Proceedings of the 2019 IEEE International Conference on Systems, Man and Cybernetics, Bari, Italy, 6–9 October 2019; pp. 2773–2778.
16. Choi, J.W.; Huh, S.; Jo, S. Improving performance in motor imagery BCI-based control applications via virtually embodied feedback. *Comput. Biol. Med.* **2020**, *127*, 104079. [CrossRef]
17. Gharabaghi, A. What Turns Assistive into Restorative Brain-Machine Interfaces? *Front. Neurosci.* **2016**, *10*, 456. [CrossRef]
18. Torkamani-Azar, M.; Kanik, S.D.; Aydin, S.; Cetin, M. Prediction of reaction time and vigilance variability from spatiospectral features of resting-state EEG in a long sustained attention task. *IEEE J. Biomed. Health Inform.* **2020**, *24*, 2550–2558. [CrossRef]
19. Ferrero, L.; Quiles, V.; Ortiz, M.; Iáñez, E.; Azorín, J.M. BCI Based on Lower-Limb Motor Imagery and a State Machine for Walking on a Treadmill. In Proceedings of the International IEEE EMBS Conference on Neural Engineering, Sorrento, Italy, 4–6 May 2020.
20. Costa, Á.; Iáñez, E.; Úbeda, A.; Hortal, E.; Del-Ama, A.J.; Gil-Agudo, Á.; Azorín, J.M. Decoding the Attentional Demands of Gait through EEG Gamma Band Features. *PLoS ONE* **2016**, *11*, e0154136. [CrossRef]
21. McFarland, D.J.; McCane, L.M.; David, S.V.; Wolpaw, J.R. Spatial filter selection for EEG-based communication. *Electroencephalogr. Clin. Neurophysiol.* **1997**, *103*, 386–394. [CrossRef]
22. Ramoser, H.; Muller-Gerking, J.; Pfurtscheller, G. Optimal spatial filtering of single trial EEG during imagined hand movement. *IEEE Trans. Rehabil. Eng.* **2000**, *8*, 441–446. [CrossRef]
23. Rainford, B.D.; Daniell, G.J. μSR frequency spectra using the maximum entropy method. *Hyperfine Interact.* **1994**, *87*, 1129–1134. [CrossRef]
24. Izenman, A. Linear Discriminant Analysis. In *Modern Multivariate Statistical Techniques*; Springer Texts in Statistics; Springer: New York, NY, USA, 2006.
25. Zich, C.; De Vos, M.; Kranczioch, C.; Debener, S. Wireless EEG with individualized channel layout enables efficient motor imagery training. *Clin. Neurophysiol.* **2015**, *126*, 698–710. [CrossRef]
26. Gordleeva, S.Y.; Lukoyanov, M.V.; Mineev, S.A.; Khoruzhko, M.A.; Mironov, V.I.; Kaplan, A.Y.; Kazantsev, V.B. Exoskeleton control system based on motor-imaginary brain-computer interface. *Sovrem. Tehnol. Med.* **2017**, *9*, 31–36. [CrossRef]

Article

Exoscarne: Assistive Strategies for an Industrial Meat Cutting System Based on Physical Human-Robot Interaction

Harsh Maithani [1,*], Juan Antonio Corrales Ramon [2], Laurent Lequievre [1], Youcef Mezouar [1] and Matthieu Alric [3]

[1] Université Clermont Auvergne, CNRS, Clermont Auvergne INP, Institut Pascal, F-63000 Clermont-Ferrand, France; laurent.lequievre@uca.fr (L.L.); youcef.mezouar@sigma-clermont.fr (Y.M.)

[2] Centro Singular de Investigación en Tecnoloxías Intelixentes (CiTIUS), Universidade de Santiago de Compostela, 15782 Santiago de Compostela, Spain; juanantonio.corrales@usc.es

[3] ADIV, ZAC des Gravanches, 10 Rue Jacqueline Auriol, 63100 Clermont-Ferrand, France; matthieu.alric@adiv.fr

* Correspondence: harshmaithani09@gmail.com

Abstract: Musculoskeletal disorders of the wrist are common in the meat industry. A proof of concept of a physical human-robot interaction (pHRI)-based assistive strategy for an industrial meat cutting system is demonstrated which can be transferred to an exoskeleton later. We discuss how a robot can assist a human in pHRI, specifically in the context of an industrial project i.e for the meat cutting industry. We developed an impedance control-based system that enables a KUKA LWR robot to provide assistive forces to a professional butcher while simultaneously allowing motion of the knife (tool) in all degrees of freedom. We developed two assistive strategies—a force amplification strategy and an intent prediction strategy—and integrated them into an impedance controller.

Keywords: physical human-robot interaction; assistive robotics; collaborative robots

Citation: Maithani, H.; Corrales, J.A.; Lequievre, L.; Mezouar, Y.; Alric, M. Exoscarne: Assistive Strategies for an Industrial Meat Cutting System Based on Physical Human-Robot Interaction. *Appl. Sci.* **2021**, *11*, 3907. https://doi.org/10.3390/app11093907

Academic Editor: Emanuele Carpanzano

Received: 28 March 2021
Accepted: 20 April 2021
Published: 26 April 2021

Publisher's Note: MDPI stays neutral with regard to jurisdictional claims in published maps and institutional affiliations.

1. Introduction and Motivation

The agri-food industry, and particularly the meat industry, is one of the most dangerous industries when it comes to employee safety. Among the various occupations, slaughtering, cutting and meat processing operations require specific dexterity to handle sharp tools or dangerous machines. However, they also require physical strength to carry heavy loads such as pallets or carcasses, quarters or muscles of meat, or to perform deboning work tasks and cutting quarters into pieces of meat. Similarly, they require performing repetitive movements or working in a cold refrigerated room and humid environment. In fact, accidents at work are common and can occur at any time. Thus, the rate of these accidents and their frequency are among the highest, among all professions combined. In addition to work-related accidents, musculoskeletal disorders (MSDs) accounted for almost 91% of occupational illness cases in 2019 [1], with 842,490 days of temporary interruption of work for all sectors. A study carried out in the Brittany region of France showed that the agri-food industries are the most risky sectors in terms of MSDs [1]. For instance, at the French level, 30% of the declared MSDs are recorded in the meat sector.

Not only is this a major problem for employees, it is also a big problem for companies and society at large. In 2019, compensation for MSDs generated two billion euros in fees (social security estimate) in France, with an average of more than 21,000 euros for each MSD stoppage, not counting the daily allowances [2]. In some companies with 10 to 20% absenteeism, MSDs disrupt production and generate additional costs. In the meat sector, MSDs involve personnel at all stages. Thus, operators involved in meat cutting, represented by boners, parers and slicers, as well as those located at manufacturing and packaging stations are affected [3]. MSDs affecting the wrist, hand and fingers represent approximately 50% of all MSDs in the meat industry.

The arduousness of tasks related to physical effort, the repeatability of movements and the agri-food environment (cold, humidity and hygiene) encountered in the meat sector deters the recruitment of young people, and ultimately few people, trained initially for this sector, remain there. This difficulty in recruiting or retaining young people leads to a significant shortage of manpower in these sectors. Technological evolution of certain workstations which would allow on the one hand a reduction of the arduousness, opening to the women certain activities which were until now reserved for the men, and on the other hand a revaluation of the trades, could change the outlook of the individuals in this sector of activity, give a positive image and promote its attractiveness.

Faced with these findings, the meat sector could soon integrate assistive and cobotic equipments to encourage companies to improve the quality of life at work of their employees. Cobotics, also known as collaborative robotics, is a technology that uses robotics, mechanics, electronics, and cognitive science to assist humans in their tasks. For the meat sector, the advantages of having cobotics are numerous:

1. Reduction of occupational risks by reducing the arduousness of the operators' tasks and improving their working conditions.
2. Overcome a major shortage of skilled labour by reducing the lack of image and attractiveness, particularly among young people (hardship in low-value jobs, etc.), which will help maintain or even develop jobs in the sector.
3. Improve the competitiveness of companies by reducing the direct and indirect costs of work stoppages and by increasing productivity.
4. Improve the safety of products by reducing the direct handling of products by operators or by integrating cleaning systems (e.g., sterilization of tools online between each operation).

In this paper, we propose the development of a proof-of-concept assistive strategy implemented through a collaborative robot in meat cutting tasks in order to reduce the musculoskeletal disorders on the wrist of human operators working in the meat industry. The developed impedance control strategy enables a KUKA LWR robot to provide assistive forces to a professional butcher while simultaneously allowing motion of the knife (tool) in all degrees of freedom. Previous robotic systems for autonomous meat handling [4] required one or several robots for performing very specific meat cutting operations. For instance, the ARMS system [5–7] was based on the separation of beef shoulder muscles, the GRIBBOT system [8] was applied to chicken breast fillet harvesting while the DEXDEB system [9,10] was useful for ham deboning. Therefore these robotic systems could not be reused for other meat handling tasks since the quality of their cut was not enough for other types of meat pieces. The new proposed robotic system (called Exoscarne) works on the principle of pHRI (physical Human-Robot Interaction) to solve this lack of generalization. From one side, the expertise of the skilled butcher is kept since the cutting trajectory is defined by the human operator, who is holding the tool at the same time as the robot. From the other side, the robot carries the load of the tool and increases the cutting force when touching the meat so that the effort applied by the human is smaller. This system results in a greater flexibility for different meat cutting tasks and can adapt itself to on-the-fly decisions made by the user.

Therefore, our new Exoscarne system, which will be described in the next sections (see Section 3 for its software components and Section 4 for its hardware components), is able to:

1. Compensate the weight of the cutting tool.
2. Permit the user to move the tool in all 6 degrees of freedom.
3. Permit impedance shaping according to the operation being performed by the user.
4. Provide assistive forces during the meat cutting operations as per the user's convenience.
5. Allow the user to perform the operations autonomously.

2. Background on Robot Assistance

One of the primary motivations of using robots for pHRI is their ability to share physical loads with their human partners. When the physical load of an object is shared or when the object is manipulated for a task, a natural division of effort between the human and the robot occurs. For example, load sharing can be for human-robot cooperative manipulation [11] or during rehabilitation [12]. Mortl et al. [13] proposed effort sharing policies for load sharing of an object by a human and multiple robots.

Some researchers focused on the larger question of selection of an 'assistance strategy' for a pHRI task. Dumora et al. [14] considered large object manipulation tasks in pHRI and proposed a library of robot assistances. Medina et al. [15] proposed a dynamic strategy selection between model-based and model-free strategies. The strategy selection is based on the concept of disagreement between the human and the robot, which in turn depends on the interaction force.

In the literature, the only example of load sharing of tool for a cooperative pHRI task, similar to meat cutting, was in [16] in which the author used a robot for assistive welding by supporting the weight of the welding equipment. In fact, most existing IADs, "Intelligent Assist devices" (i.e., active cobotics systems for human assistance) [17], are used in the automotive industry for the quasi-static collaborative transportation of heavy loads (e.g., motors, doors ...). However, these solutions are not suitable for the meat industry [18] since they can only assist through specific directions in the work-space [19] (while meat cutting requires complex 6D trajectories), they do not integrate safety solutions for handling dangerous tools (such as the knife for meat cutting) and they do not take into account the important dynamic non-linear effects of the meat cutting operations [20]. To the best of our knowledge, there is no prior pHRI related work which uses a robot for assisting a human for cutting meat by taking into account not only a classical force amplification strategy (such as in [20]) but also an intent prediction module in order to reduce the final forces to be applied by the human operator.

3. Methodology

3.1. pHRI Assistive Strategies

The main interaction controllers for pHRI are impedance and admittance control (see Section 3.2 for their mathematical definition). Controller stability issues are common with admittance control [21,22]. When a human holds the tool at the end-effector, it results in a coupled system that can lose stability if the human operator stiffens his arm muscles, leading to robot vibrations. Hence for this task we chose impedance control. Investigating the stability issues of admittance control is a field in itself and there are several heuristic methods in the literature [20]. Impedance control was possible as the robot had a torque sensor in each joint, enabling torque control. Admittance control is preferred with robots that have only position control, by using an external FT sensor. As the task involved motion in the cartesian space hence we used the cartesian impedance controller that is explained later. We devised two assistive strategies:

1. Force amplification strategy

   In this strategy we amplify the forces applied by the user on the knife's handle (see Section 4 for the experimental setup), detected by the FT sensor and input to our control scheme. The control diagram is shown in Figure 1. The appropriate parameter η has to be determined, as explained in Section 5.3.

2. Intent prediction strategy

   In this strategy we predict the forces to be applied by the user using RNN-LSTM networks explained in Section 3.3. The control diagram is shown in Figure 2.

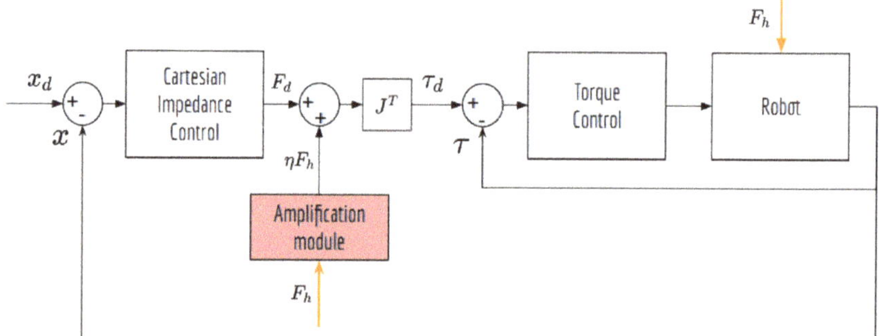

Figure 1. Impedance controller diagram with amplification module for the force amplification strategy.

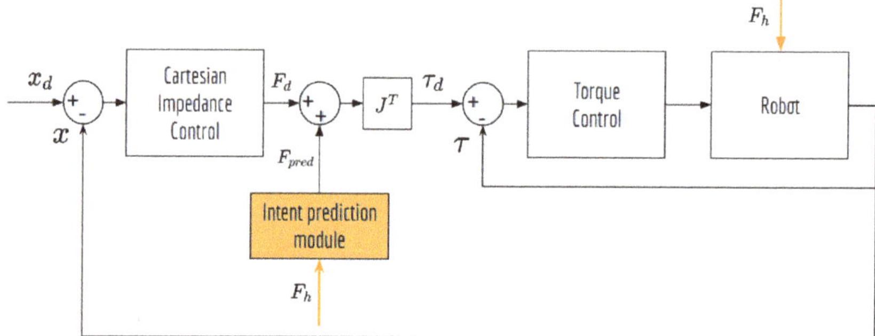

Figure 2. Impedance controller diagram with intent prediction module for intent prediction strategy.

Both Figures 1 and 2 are identical except for the module of their respective assistive strategies. In the force amplification strategy, the amplification module amplifies the human user's force input F_h at each time step through robot assistance ηF_h. In the intent prediction strategy, a trained RNN-LSTM network takes the human user's force input F_h at each time step to anticipate the user's input for the next time step F_{pred} and provides this force via robot assistance. In both cases the user can haptically sense this assistance being provided by the robot.

3.2. Impedance Control

The forward kinematics of a robotic manipulator is written as [23]:

$$x(t) = f(q) \tag{1}$$

where $x(t) \in \Re^n$ and $q \in \Re^n$ are the pose (i.e., position/orientation) of the end-effector in the Cartesian space and the joint angle coordinates in the joint space, respectively. Differential kinematics is obtained by deriving (1) with respect to time:

$$\dot{x}(t) = J(q)\dot{q} \tag{2}$$

where $J(q) \in \Re^{n \times n}$ is the Jacobian matrix. Differentiating (2) again results in the acceleration of the end-effector:

$$\ddot{x}(t) = \dot{J}(q)q + J(q)\ddot{q} \tag{3}$$

The robot arm dynamics in the joint space is given by:

$$M(q)\ddot{q} + C(q,\dot{q})\dot{q} + G(q) = \tau - J^T(q)F \tag{4}$$

where $M(q) \in \Re^{n \times n}$ is the symmetric positive-definite inertia matrix; $C(q, \dot{q})\dot{q} \in \Re^n$ is the Coriolis and Centrifugal forces; $G(q) \in \Re^n$ is the gravitational force; $\tau \in \Re^n$ is the vector of control input; $F \in \Re^n$ denotes the net force exerted by the robot on the environment at the end-effector ($F = F_r - F_{ext}$). Otherwise, $F_r \in \Re^n$ is the force exerted by the robot on the environment at the end-effector while $F_{ext} \in \Re^n$ is the external force exerted by the environment on the robot at the end-effector. In our pHRI task, the environment is the human, specifically the human hand that is in contact with the robot. The robot dynamics can be written in the Cartesian space as:

$$M_x(q)\ddot{x} + C_x(q, \dot{q})\dot{x} + G_x(q) = u - F \tag{5}$$

where

$$
\begin{aligned}
M_x(q) &= J^{-T}(q)M(q)J^{-1}(q), \\
C_x(q, \dot{q}) &= J^{-T}(q)(C(q, \dot{q}) - M(q)J^{-1}(q)\dot{J}(q))J^{-1}(q), \\
G_x(q) &= J^{-T}(q)G(q), \, u = J^{-T}(q)\tau, \, F = F_r - F_{ext}
\end{aligned}
$$

As explained in Section 3.1, the most common interaction controllers for pHRI are admittance and impedance control. Both controllers are based on a target impedance model for the robot and they only differ in terms of input and output (see Figures 3 and 4). Therefore, the target robot impedance model can be represented as a mass-damper-spring system:

$$M_d(\ddot{x}_d - \ddot{x}) + B_d(\dot{x}_d - \dot{x}) + K_d(x_d - x) = F_d = F_r \tag{6}$$

where M_d, B_d, K_d are the virtual inertia, damping and stiffness of the robot, respectively. F_d is the desired force and x_d can be interpreted as the rest position of this virtual mass-damper-spring system. For pHRI tasks in which the human touches the robot, the human limb (arm+hand) can also be modeled as a mass-damper-spring system:

$$M_h\ddot{x}_h + B_h\dot{x}_h + K_h(x_h - x_{hd}) = F_h \tag{7}$$

where M_h, B_h, K_h are the limb inertia, damping and stiffness respectively and F_h is the force applied by the human to the robot. The limb impedance values are not fixed and depend from person to person, as well as on the task being carried out. x_h is the position of the human wrist in the robot frame and x_{hd} can be interpreted as the desired target position. Discussion on the limb impedance parameters or their calculation are not in the purview of this paper. In fact, the investigation of the limb impedance parameters would have been necessitated if an admittance controller was used, to tackle the potential controller stability issues. Since we have used a torque-based impedance controller for our experiments, it does not face such issues. While real time measurements of limb impedance values could be used for interpreting the intent of the human operator during the operation, this would have required attaching EMG sensors on the arms of the human operator, such as in [24], thereby reducing the practicality of this research for an industrial meat cutting scenario. In our experiments the human intent is interpreted via the force measurements read by the FT sensor, while staying comfortable and intuitive for the human operator.

When no assistance strategy is required and we want the robot to freely follow the motion of the human, we can set $K_d = 0$ or $x = x_d$. This "direct teaching mode" is essentially based on the elimination of the spring component of the robot impedance model ("minimum impedance" strategy in Section 5.4) and avoids any restoring forces. If the human holds the tool rigidly then we can assume that the forces are transmitted completely to the robot. If the hand is close enough to the end-effector, we can assume that their positions, velocities and forces are equivalent: $x_d = x_{hd}$, $\dot{x}_h = \dot{x}$ and $\ddot{x}_h = \ddot{x}$.

In admittance control (popularly called position-based impedance control) the input is the force applied by the environment and the output is displacement/velocity as shown in Figure 3. In pHRI tasks, the force is applied by the human on the tool at the end-effector ($F_{ext} = F_h$) and it is measured by the FT sensor. The difference between the desired force for the robot ($F_d = F_r$) and this external force is applied as input to the impedance model in (6) in order to obtain the desired position for the end-effector x_d as output:

$$M_d(\ddot{x}_d - \ddot{x}) + B_d(\dot{x}_d - \dot{x}) + K_d(x_d - x) = F_d - F_h \tag{8}$$

On the contrary, the impedance control (torque-based impedance control) has as input the end-effector displacement/velocity and the desired force as output, as shown in Figure 4. In this case, the desired robot impedance model is the same as in (6) but the difference between the desired robot force and the external/human force is transformed into joint torques for the robot control.

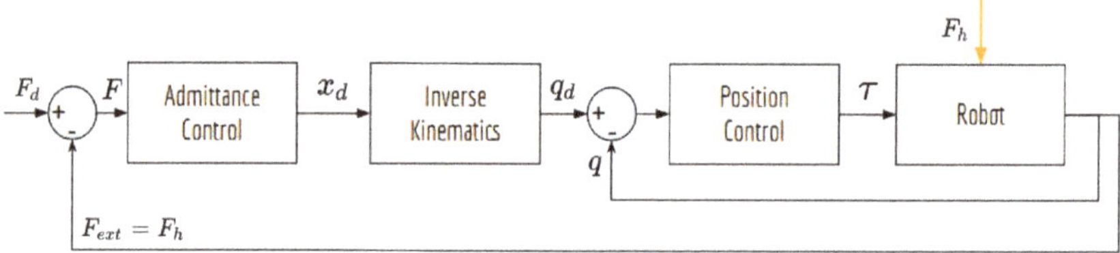

Figure 3. Admittance control (position-based impedance control).

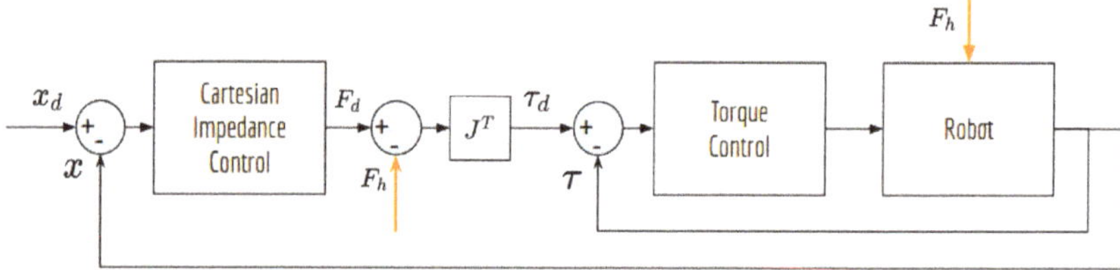

Figure 4. Torque-based impedance control.

3.3. Long Short-Term Memory Model

For the intent prediction module, we use RNN-LSTM units [25]. RNNs (Recurrent Neural Networks) are based on processing sequential data, especially temporal data as they have an internal memory. In fact, they are useful for making predictions using time-series data [23]. RNNs can be improved by using what are called LSTM units in order to solve the vanishing gradient problem (i.e., gradients tend to disappear in RNNs when backpropagating errors in too long sequences). In fact, each LSTM unit has a special structure composed by 3 gates to control what information to keep and what to forget so that more stable errors are backpropagated (see Figure 5):

1. An input gate (i).
2. An output gate (o).
3. A forget gate (f).

As a result, RNNs with LSTM units are able to learn long-term dependencies within data sequences that were not possible only with RNNs. Given an input sequence $\bar{x}_1, \bar{x}_2, ..., \bar{x}_t$ the LSTM unit maps the input sequence to a sequence of hidden states $h_1, h_2, ..., h_t$ (which are also the outputs) by passing information through a combination of gates (see Figure 6):

The Input gate (for updating the cell) is:

$$i_t = \sigma_g(W_i \bar{x}_t + R_i h_{t-1} + b_i) \tag{9}$$

The Forget gate (for reseting the cell/forgetting) is:

$$f_t = \sigma_g(W_f \bar{x}_t + R_f h_{t-1} + b_f) \tag{10}$$

The Cell candidate (for adding information to the cell) is:

$$g_t = \sigma_c(W_g \bar{x}_t + R_g h_{t-1} + b_g) \tag{11}$$

where σ_c is the state activation function (here it is the hyperbolic tangent function: $\sigma_c = tanh(\bar{x})$).

The Output gate is:

$$o_t = \sigma_g(W_o \bar{x}_t + R_o h_{t-1} + b_o) \tag{12}$$

where σ_g is the gate activation (here it is the sigmoid function: $\sigma(x) = (1 + e^{-\bar{x}})^{-1}$)

The Memory Cell state at timestep t is:

$$c_t = f_t \odot c_{t-1} + i_t \odot g_t \tag{13}$$

Here $\odot$ is the Hadamard product (element-wise multiplication of vectors). The memory cell selectively retains information from previous timesteps by controlling what to remember via the forget gate f_t.

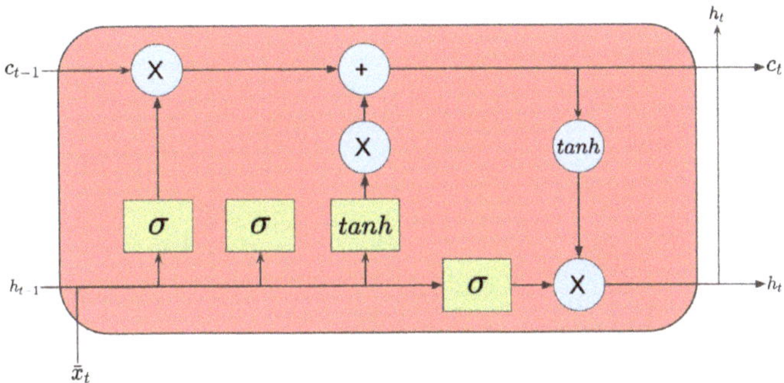

Figure 5. A single LSTM unit.

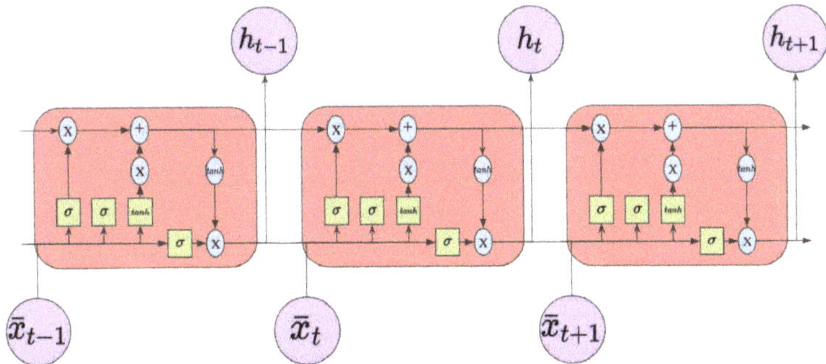

Figure 6. An unrolled Recurrent Neural Network with LSTM units.

The Hidden state (also called Output state) at time step t is :

$$h_t = o_t \odot \sigma_c(c_t) \tag{14}$$

The hidden state is passed as input to the next timestep and thus, it is possible to stack numerous LSTM units (see Figure 6). W_i, W_f, W_g are the learnable input weights; R_i, R_f, R_g are the learnable recurrent weights and b_i, b_f, b_g are the learnable bias. By using memory cells and hidden states, LSTM units are able to retain information. The sigmoid function is a good activation function for the 3 gates (In, Out and Forget) since it outputs a value between 0 and 1. However for the memory cell, the values should be able to increase or decrease (which is not possible with the sigmoid activation function whose output is always non-negative). Therefore, the hyperbolic tangent function (*tanh*) is used as the activation function for the memory cell.

4. Experimental Setup

4.1. KUKA LWR Robot

The KUKA LWR is a `Torque-controlled Flexible Robot` with 7 degrees of freedom. The design and control concepts of the robot have been discussed in [26,27]. There are two control modes available. The joint position control is implemented at a frequency of 3 kHz (decentralized control) [28]. Inverse kinematics and the cartesian impedance control mode run at a frequency of 1 kHz. The KUKA LWR has a torque sensor at each joint that enables torque control and impedance control. It also has motor side position sensors, as well as link side position sensors. Due to friction it is difficult for robots to implement torque control only with motor current commands [28].

The default coordinate system of the KUKA LWR is `Right Handed System` for which `det(R)` = 1. The dimensions of the robot can be retrieved from the official manufacturer documentation. Using these dimensions we can obtain the DH parameters (refer Table 1). To simulate the robot, an alternate form of representation called URDF is shown in Table 2 and its corresponding 3D model is shown in Figure 7. The Unified Robotic Description Format (URDF) is an XML file format used in ROS to describe all elements of a robot. The robot interacts with the external PC through the `Fast Research Interface` [29] through three modes—(a) Joint position control (b) Joint impedance control mode and (c) Cartesian impedance control mode.

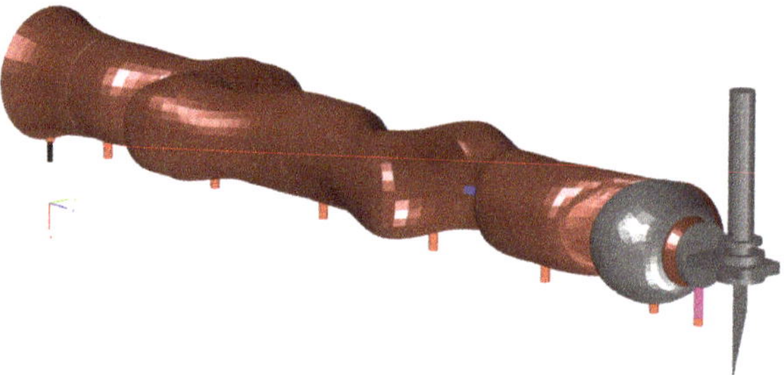

Figure 7. Exoscarne 3D simulated model with the KUKA LWR arm in Home configuration.

Table 1. DH parameters of KUKA LWR 4+ robot.

	Joints	d_i(m)	q_i(rad)	a_i	α_i(rad)	q_{min}	q_{max}	τ_{max}(Nm)
J1	A1	0.3105	q_1	0	$\pi/2$	-170	170	176
J2	A2	0	q_2	0	$-\pi/2$	-120	120	176
J3	E1	0.4	q_3	0	$-\pi/2$	-170	170	100
J4	A3	0	q_4	0	$\pi/2$	-120	120	100
J5	A4	0.39	q_5	0	$\pi/2$	-170	170	100
J6	A5	0	q_6	0	$-\pi/2$	-120	120	38
J7	A6	0.078	q_7	0	0	-170	170	38

Table 2. URDF description of KUKA LWR 4+ robot.

Joints	x(m)	y(m)	z(m)	r	p	y	Axis	
A1	0	0	0.11	0	0	0	[0 0 1]	+z axis
A2	0	0	0.2005	0	0	0	[0 −1 0]	−y axis
E1	0	0	0.2	0	0	0	[0 0 1]	+z axis
A3	0	0	0.2	0	0	0	[0 1 0]	+y axis
A4	0	0	0.2	0	0	0	[0 0 1]	+z axis
A5	0	0	0.19	0	0	0	[0 −1 0]	−y axis
A6	0	0	0.078	0	0	0	[0 0 1]	+z axis

To fulfill our objectives we used the cartesian impedance control mode available with the KUKA LWR robot. The KUKA manual [30] states that the control law for the cartesian impedance controller is

$$\tau_{cmd} = J^T \left(K_c (x_{desired} - x_{current}) + F_{cmd} \right) + D(d_c) + f_{dynamics}(q, \dot{q}, \ddot{q}) \tag{15}$$

where $q \in \Re^n$ is the joint position vector, K_d is the stiffness matrix in the end-effector frame, D_d is the normalized damping parameter in the end-effector frame, x and x_d are the current and the desired pose of the end-effector respectively in the global frame. The translational stiffness $K_x, K_y, K_z \in [0.01, 5000]$ N/m and rotational stiffness $K_{A_z}, K_{B_y}, K_{C_x} \in [0.01, 300]$ N/m-rad.

The KUKA LWR has an inbuilt external tool calibration functionality. Using this feature the robot can account for external tool dynamics as well, thereby enabling gravity compensation for the tool.

In this work we used the KUKA Fast Research Interface (FRI [29]), ROS [31], Kinematics and Dynamics library (KDL [32]) , the MATLAB toolbox by Peter Corke (RCV [33]), and the MATLAB Robotics Toolbox (RTB [34]). For the intent prediction strategy, the high level program was written in MATLAB and connected to the network via ROS (Robot Operating System), by using the MATLAB Robotics Toolbox.

4.2. ATI FT Sensors

In order for the robot to provide assistive forces as per user comfort we had to measure how much forces are being applied by the user. For this, we used two 6-axis ATI Gamma force-torque sensors (see Figure 8): one mounted below the joystick (sensor B) and another mounted on the end-effector of the arm (sensor A, see Figure 9). Both FT (Force-Torque) sensors provide a 6-dimensional wrench in the sensor frame at 1000 Hz.

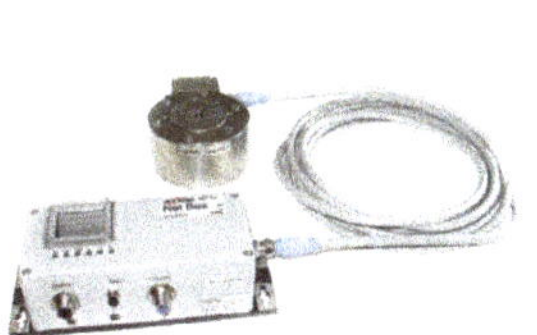

(a) FT ATI acquisition system

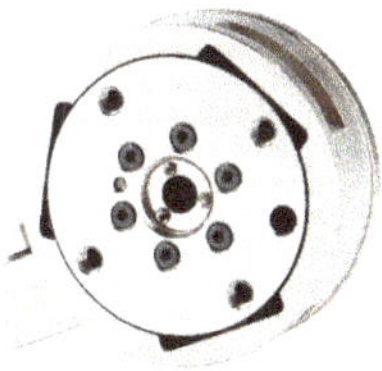

(b) FT sensor

(c) Internal view

(d) Internal view 2

Figure 8. Photos of the force-torque (FT) sensor.

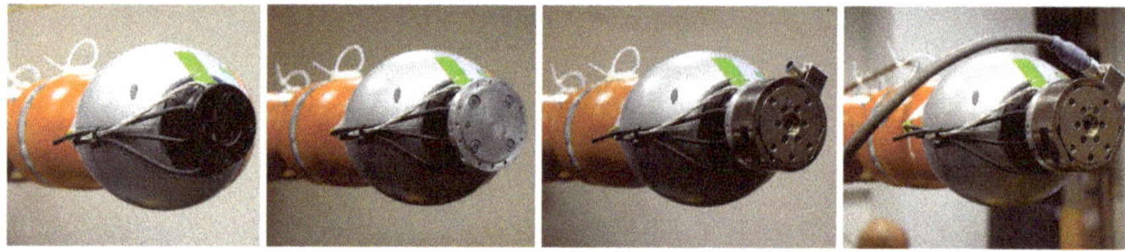

Figure 9. Attaching the FT sensor A to the KUKA LWR 4+ robot end-effector.

The relation between the different frames associated to these sensors (A and B), to the tool (i.e., tool center point—TCP—or tip of the knife) and the robot world frame O are shown in Table 3 and visualized in Figure 10. To use the impedance controller all the forces must be in the same frame. This requires transformation of the sensed forces in the sensor frame to the end-effector frame of the robot. The equation for transformation of forces from one frame to another is:

$$\begin{bmatrix} ^A F_A \\ ^A M_A \end{bmatrix} = \begin{bmatrix} ^B_A R & 0 \\ [^A t_x]^B_A R & ^B_A R \end{bmatrix} \begin{bmatrix} ^B F_B \\ ^B M_B \end{bmatrix} \tag{16}$$

where $[t_x] = \begin{bmatrix} 0 & -t_z & t_y \\ t_z & 0 & -t_x \\ -t_y & t_x & 0 \end{bmatrix}$ or

$$^A F_A = {}^B_A T_f \, {}^B F_B \tag{17}$$

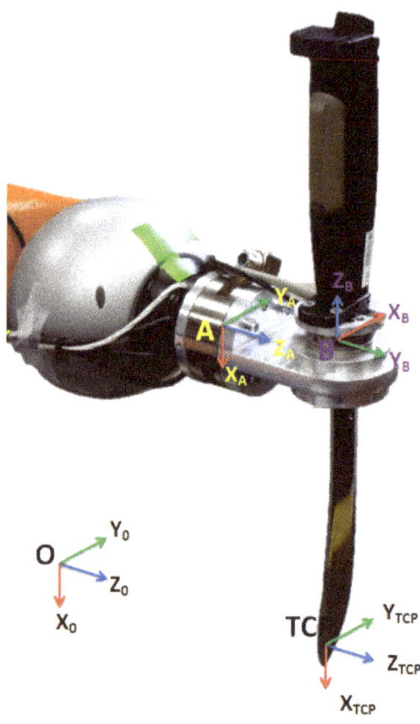

Figure 10. Visualization of all frames relating the tool -knife- with the FT sensors A/B.

Table 3. Joystick frames when the robot is in Home configuration. Sensor A is always aligned with the robot end-effector frame.

Robot World Frame	Sensor A (FT13855)	Sensor B (FT13953)	Relation
X_o	X_A	$-Z_B$	$X_o \,\|\, X_A \,\|\, -Z_B$
Y_o	Y_A	X_B	$Y_o \,\|\, X_A \,\|\, X_B$
Z_o	Z_A	Y_B	$Z_o \,\|\, Z_A \,\|\, Y_B$

4.3. Allen-Bradley Joystick

The joystick (i.e., the knife handle) is an Allen-Bradley 440J-N enabling switch. It has 8 electrical connections and when the joystick switch is pushed or released, combinations of these 8 connections are activated as shown in Figure 11.

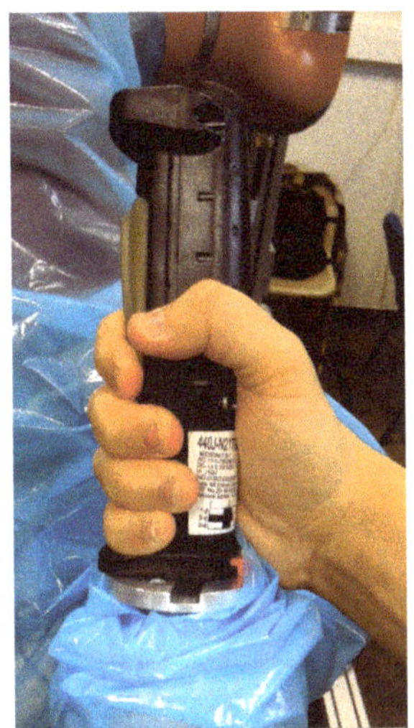

(a) Joystick pressed

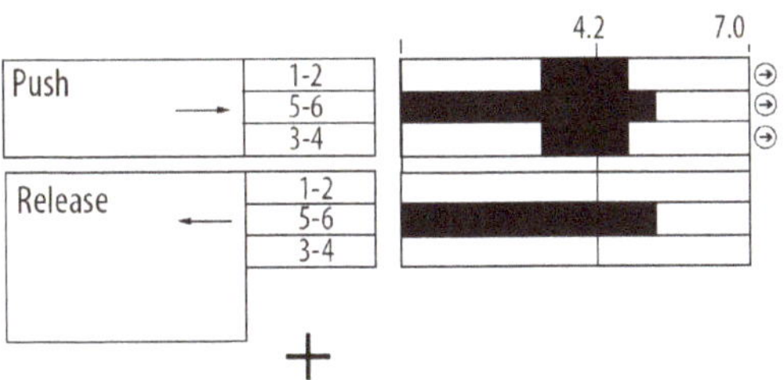

(b) Electrical configuration of the 2 joystick buttons. Above: lateral dead man's switch (1–6 contacts). Below: upper push-button (7–8 contacts).

Figure 11. Allen-Bradley joystick used as the handle of the knife.

The 8 electrical wires from the joystick are connected to an Arduino Uno circuit board which was then connected to a laptop (see Figure 12 for the complete connection diagram). The `rosserial` ROS package and the `ros_lib` library is used to integrate the Arduino Uno with the robot network via ROS.

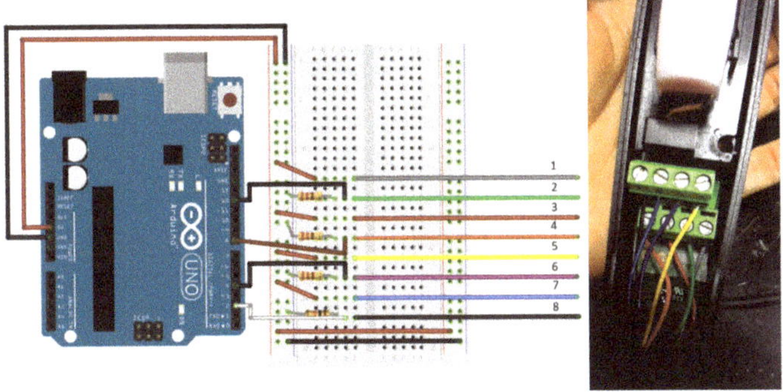

Figure 12. Arduino circuit diagram for connecting the joystick to a laptop.

By using this joystick, the user will be able to communicate his intention clearly with regards to the meat cutting operation. See Section 5.2 for a more detailed description of how the user will use this joystick for a safe and unobtrusive physical human-robot interaction.

4.4. Meat Cutting Equipment

To fully understand the practical issues in implementing the technical aspects of the project we first performed some meat cutting trials using the robot in the gravity compensation mode. A special holding tool was developed to accommodate a joystick which would be held by the user and the cutting tool. The project involved meat operations on chicken, pork and beef for which special cutting tools were made. The joystick was an industrial joystick which we customized for the project and it allowed a very natural and intuitive user interface. The blades are professional butcher blades customized for the project by machining in the laboratory. A customized pneumatic hanger was developed for holding the pork vertically. Specific cutting tools and holding apparatus for each meat product or carcass to cut were devised as shown in Figure 13 below.

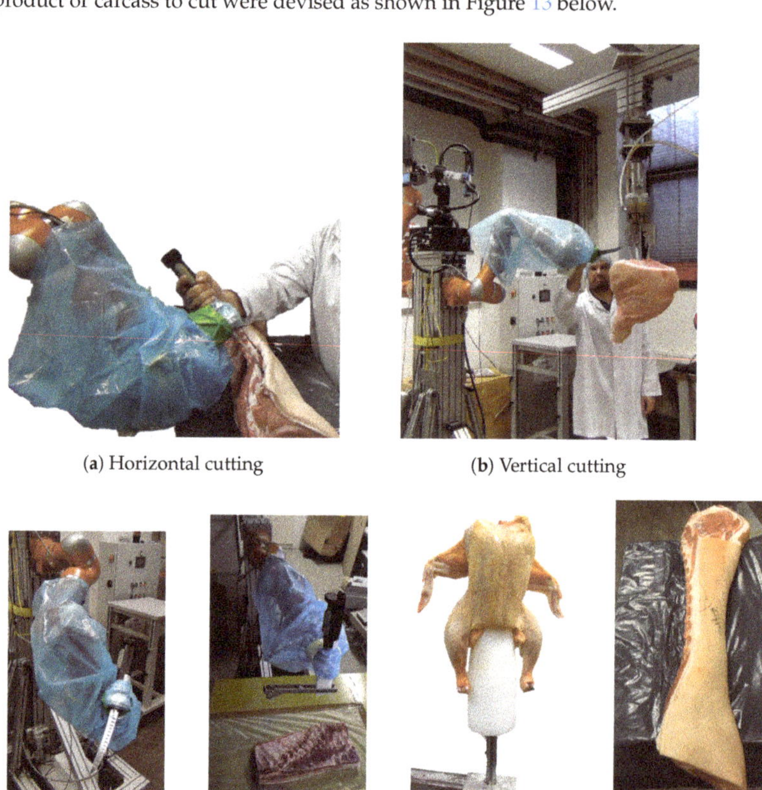

(**a**) Horizontal cutting   (**b**) Vertical cutting

(**c**) Knife   (**d**) Peeler   (**e**) Chicken holder   (**f**) Pork

Figure 13. Exoscarne equipment.

5. Experiments

The complete network diagram is shown in Figure 14. All the PCs of this network communicate via ROS. The two assistive strategies are implemented in the right laptop, which recovers FT sensor information from the two middle PCs and sends robot commands to the left PC.

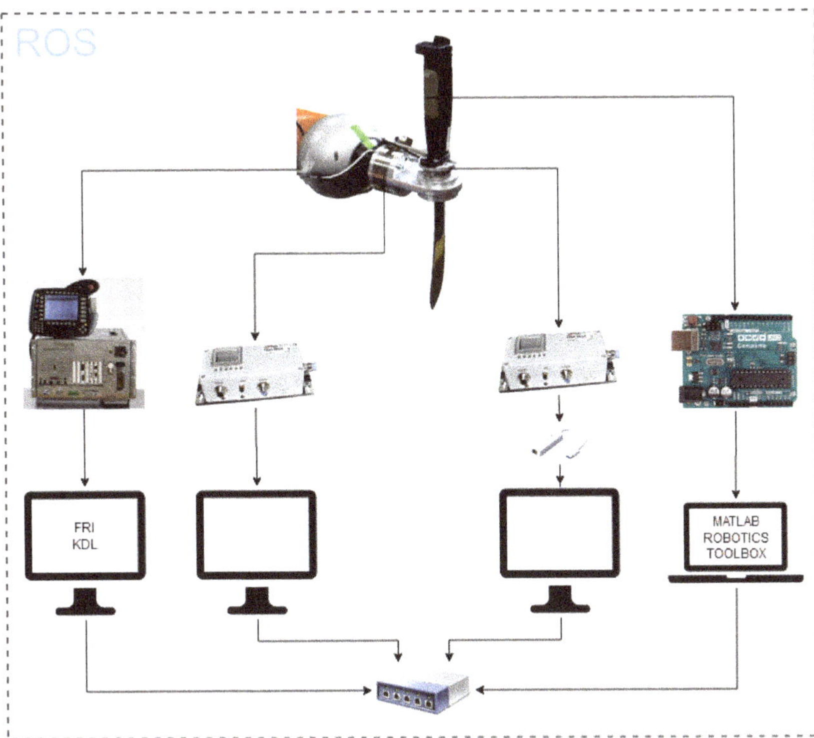

Figure 14. Network diagram of the Exoscarne system. From left to right: control box of the KUKA LWR 4+ robot connected to one PC with FRI/KDL libraries; data acquisition boxes for FT sensors A and B connected to two PCs and an Arduino board for joystick acquisition connected to a laptop.

We performed experiments to address the following questions:

1. Which FT sensor should be used as a source of input for the control scheme?
2. What should the impedance shaping strategy performed through the joystick buttons be?
3. What should the amplification factor be for the force amplification strategy?
4. Comparison of the force amplification strategy and the intent prediction strategy.

5.1. Comparison of FT Sensors

As shown in Figure 10, there were two FT sensors on the cutting tool. We wanted to confirm if 2 FT sensors provide an advantage over a single FT sensor, as well as determine which FT sensor is better as a source of input for the control scheme.

A piece of thick foam was used as the material to be cut. This foam reproduces the same type of shearing forces as meat cutting. Two experiments were performed. In the first one the robot was commanded to apply forces gradually from 0 to 50 N in the Y-direction of the world frame, which was parallel to the cutting direction of the knife. In the second experiment, a human cut the foam multiple times in the Y-direction, as shown in Figure 15.

Based on the sensor readings from both the sensors, it was observed that for free motion in space (without any cutting), both the sensors record the same readings. However, for cutting motion, sensor A (refer to Figure 10) is blind to the cutting forces. Sensor B recorded forces when the robot cut the foam autonomously, but the recorded forces were feeble and almost negligible to the forces commanded to the robot. When a human cut the foam using impedance controller the sensor B registered adequate forces.

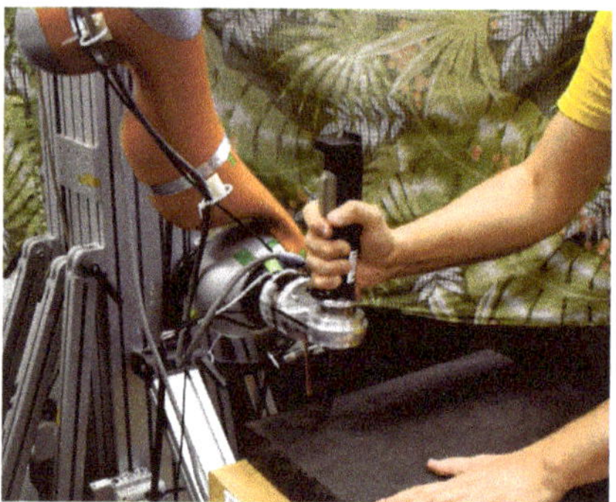

Figure 15. Comparison of sensor A and sensor B.

Hence it was determined that it is reasonable to consider that the forces sensed by sensor B (below the joystick) capture the human applied forces on the knife and thus, the user's intention. During actual meat cutting experiments, both the sensors were alternated as sensor inputs for the controller in numerous trials and the user responded that he preferred the system behaviour when sensor B was the sensor input.

5.2. Impedance Shaping for Cutting

While cutting meat, the user performs an active meat cutting operation followed by an inactive repositioning of the knife for the next stroke. Furthermore, the user may want to use both of his hands for some other task and as such need the robot to stay still in the last position. The joystick has two buttons (see Figure 10): the first lateral grey button with three possible positions (i.e., a dead-man's switch) and the second upper black button having two states (i.e., a pushbutton). This gives us a total of 6 possible combinations shown in Table 4. The corresponding electrical connections of both buttons are shown in Figure 11.

Table 4. States of joystick buttons.

		Button 2	
		State 0	State 1
Button 1	Position 0	Robot is stiff + no amplification	Robot is stiff + amplification
	Position 1	Robot is free, but no amplification (used for positioning of knife)	Robot is free + amplification (used for cutting)
	Position 2	Robot is free, but no amplification (used for positioning of knife)	Robot is free + amplification (used for cutting)

The button configurations from Table 4 are interpreted in the impedance shaping algorithm as shown in c. We can see that the last two rows are identical. Originally we tried to use position 2 with an impedance relation (i.e., $K, D \leftarrow F_h$); however, the user preferred to keep the interface simple and make the position 1 and position 2 identical for operation. This shows that developing user-friendly interface is essential for the adoption of pHRI over human-only or robot-only equipment. The constant values of Table 5 are:

$K_{min} = \begin{bmatrix} 0.1 & 0.1 & 0.1 & 0.1 & 0.1 & 0.1 \end{bmatrix}$,
$K_{max} = \begin{bmatrix} 5000 & 5000 & 5000 & 300 & 300 & 300 \end{bmatrix}$,
$D_{min} = \begin{bmatrix} 0.01 & 0.01 & 0.01 & 0.01 & 0.01 & 0.01 \end{bmatrix}$,
$D_{max} = \begin{bmatrix} 1.0 & 1.0 & 1.0 & 1.0 & 1.0 & 1.0 \end{bmatrix}$

Table 5. Impedance shaping using joystick.

		Button 2	
		State 0	State 1
Button 1	Position 0	$K_{max}, D_{max}, \eta = 0$	K_{max}, D_{max}, η
	Position 1	$K_{min}, D_{min}, \eta = 0$ (used for positioning of knife)	K_{min}, D_{min}, η (used for cutting)
	Position 2	$K_{min}, D_{min}, \eta = 0$ (used for positioning of knife)	K_{min}, D_{min}, η (used for cutting)

5.3. Tuning the Amplification Factor

In the force amplification strategy we amplify the forces applied by the user on the joystick, detected by the FT sensor and input to our control scheme (see Figure 1). For the robot to provide assistive forces we had to determine the amplification factor η for each degree of freedom as shown in Equation (18):

$$F_{cmd} = \begin{bmatrix} {}^eF_x \\ {}^eF_y \\ {}^eF_z \\ {}^eF_{A_z} \\ {}^eF_{B_y} \\ {}^eF_{C_x} \end{bmatrix} = \begin{bmatrix} \eta_x \\ \eta_y \\ \eta_z \\ \eta_{A_z} \\ \eta_{B_y} \\ \eta_{C_x} \end{bmatrix} \begin{bmatrix} {}^e_B F_x \\ {}^e_B F_y \\ {}^e_B F_z \\ {}^e_B \tau_z \\ {}^e_B \tau_y \\ {}^e_B \tau_x \end{bmatrix} \tag{18}$$

While a high amplification factor would easen the load on the user, it could also give the user the perception that he is no longer in control of the operation and reduce his comfort with the system (as was realized during the experiments). This is the reason we cannot have the robot simply apply the highest forces possible.

The tuning of the comfortable amplification factors was a continuous process where several iterations of meat cutting were performed by a professional butcher. It was decided collectively that there would be only two amplification factors: one common η_f for forces and one η_τ for torques (i.e., $\eta_f = \eta_x = \eta_y = \eta_z$ and $\eta_\tau = \eta_{A_z} = \eta_{B_y} = \eta_{C_x}$). This iterative tuning process finished when the butcher found that the assistance behavior was comfortable, as explained in Figure 16.

The meat cutting operation involves both the application of forces and torques, as such the force amplification factor η_f cannot be determined independently of the torque amplification factor η_τ. In earlier experiments the user felt that $\eta_\tau = 3$ was ideal for him and hence to determine η_f a series of consecutive experiments were done as shown in Table 6 enabling the user to make a subjective comparison.

At the end of the experiment it was concluded that the user preferred $\eta_f \in [10, 20]$ and $\eta_\tau = 3$. With $\eta_f = 20$ and $\eta_\tau = 4$, the user found the system to be too reactive and he felt he was no longer in control. These 6 experiments and the temporal evolution of the cutting forces applied by the user after applying these optimal force amplification factors are shown in the next section.

5.4. Meat Cutting with Force Amplification Strategy

As a knife mounted on the end-effector of a robot is inherently dangerous, the knife for the meat cutting experiments was always covered with a sheath when not in use to avoid accidental injury. When experiments were being performed, even if the user could stop the robot at any time by releasing the button, a second person always had his/her hand on the emergency robot switch to disable the robot if something goes wrong. The user also wore a protective gear around his arms. In addition, as a precaution, the robot was wrapped with a plastic sheet to prevent minute meat pieces from entering the internal structure of the robot.

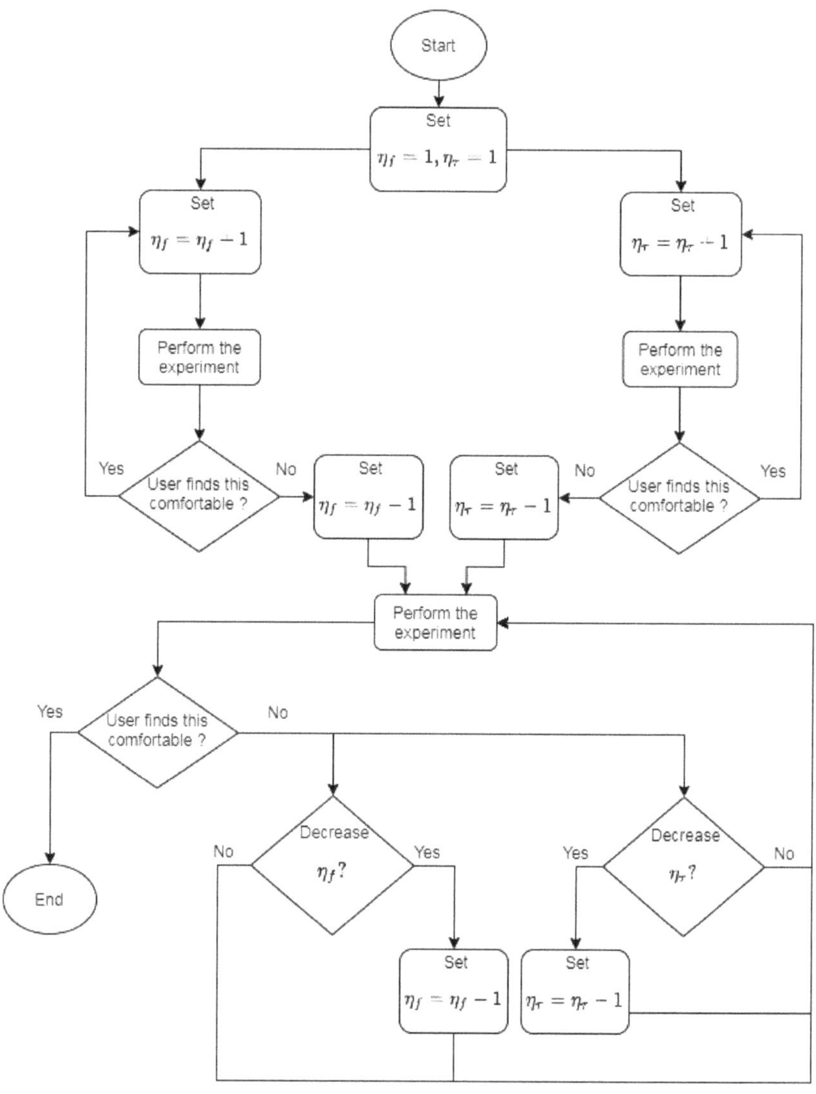

Figure 16. Iterative procedure for tuning the two amplification factors for forces and torques.

Table 6. Determining the force amplification factor η_F.

S.No.	Meat	Part	Side	Operation	FT Sensor Input	Force Amplification η_f	Torque Amplification η_τ
1	Pork	Square	Right	Cutting	B	20	3
2	Pork	Square	Right	Cutting	B	20	3
3	Pork	Square	Right	Cutting	B	15	3
4	Pork	Square	Right	Cutting	B	15	3
5	Pork	Square	Right	Cutting	B	10	3
6	Pork	Square	Right	Cutting	B	20	3

Experiment 1—Cobot assisted pork cutting

Figures 17 and 18 show the sequence of cuts of Experiment 1 and the corresponding temporal evolution of the XYZ-total forces applied by the user and measured by the B sensor, respectively. The terminology A cut, B cut, etc. in these figures are author defined to refer to the iteration of the cutting and not the cutting methods. Therefore, the images shown in Figure 17 do not represent one single cut in progress, but instead one image of each cut: they can be interpreted as cut 1, cut 2, etc.

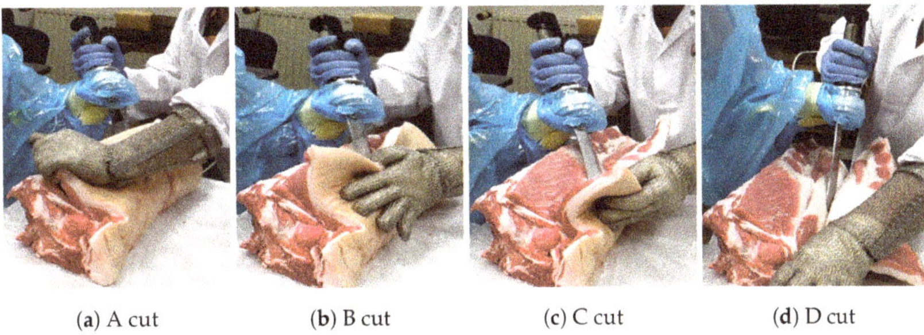

(**a**) A cut   (**b**) B cut   (**c**) C cut   (**d**) D cut

Figure 17. Experiment 1—cobot assisted pork cutting.

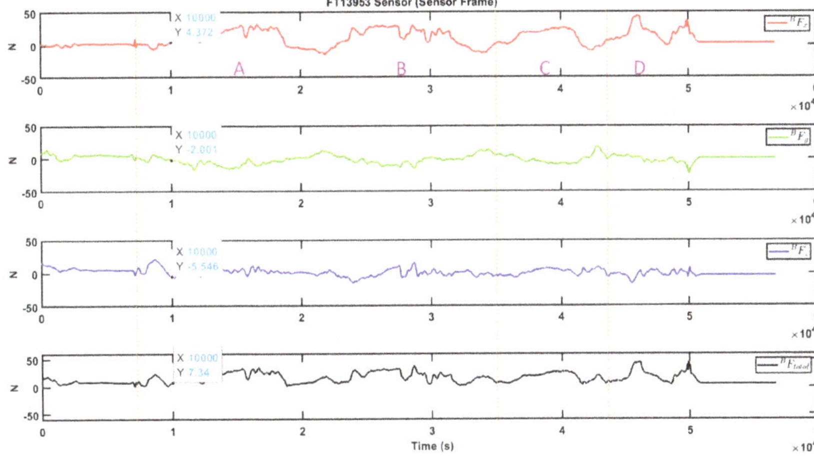

Figure 18. Forces applied by the user in Experiment 1.

Experiment 2—Cobot assisted pork cutting

In Experiment 1, it took the butcher (the user) 4 cuts to make 1 slice. Nevertheless, in Experiment 2 (see Figure 19), with the same section but from a different meat, it took him 9 cuts to make 1 slice (see Figure 20 for the corresponding temporal force evolution). This is due to the natural variation in the body composition from one animal meat to another and also how much forces the user wanted to apply for a single cut.

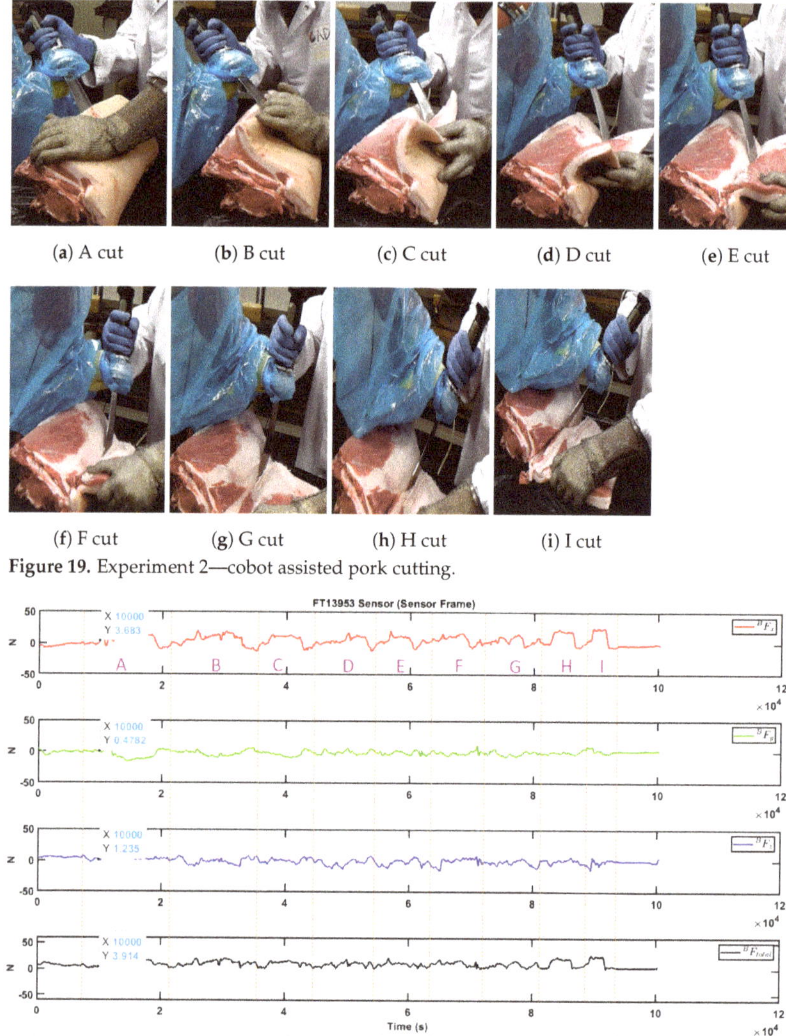

(**a**) A cut (**b**) B cut (**c**) C cut (**d**) D cut (**e**) E cut

(**f**) F cut (**g**) G cut (**h**) H cut (**i**) I cut

Figure 19. Experiment 2—cobot assisted pork cutting.

Figure 20. Forces applied by the user in Experiment 2.

Experiment 3—Cobot assisted pork cutting

Figure 21 shows the 5 cuts of Experiment 3 while Figure 22 represents the temporal evolution of the forces exerted by the human during those cuts while assisted by the force amplification strategy.

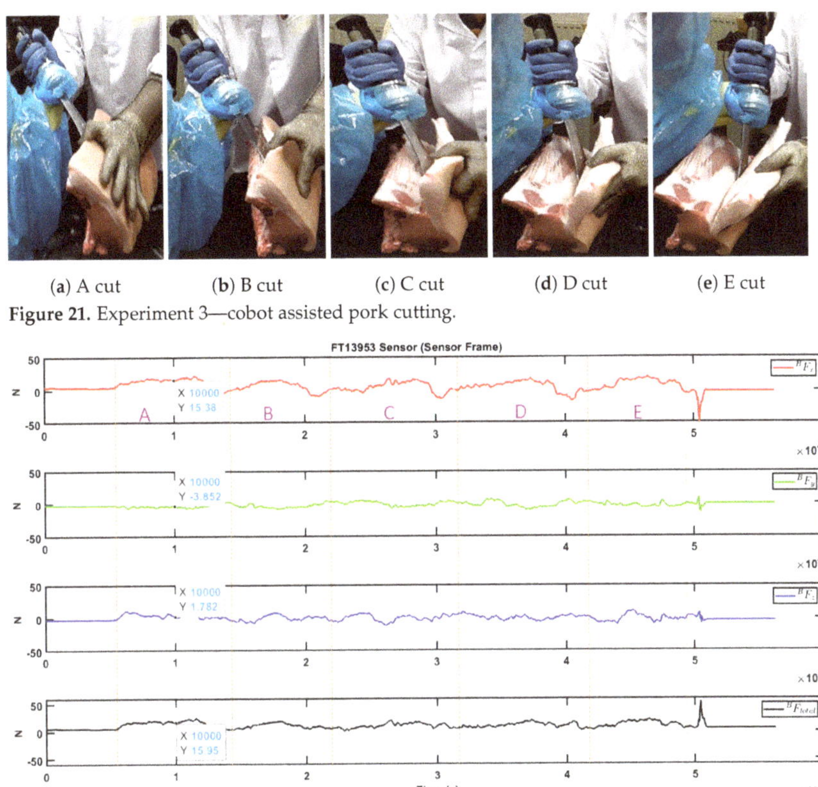

(**a**) A cut   (**b**) B cut   (**c**) C cut   (**d**) D cut   (**e**) E cut

Figure 21. Experiment 3—cobot assisted pork cutting.

Figure 22. Forces applied by the user in Experiment 3.

Experiment 4—Manual pork cutting

Figure 23 shows the 9 cuts of Experiment 4 and Figure 24 represents the corresponding temporal evolution of the forces applied by the human.

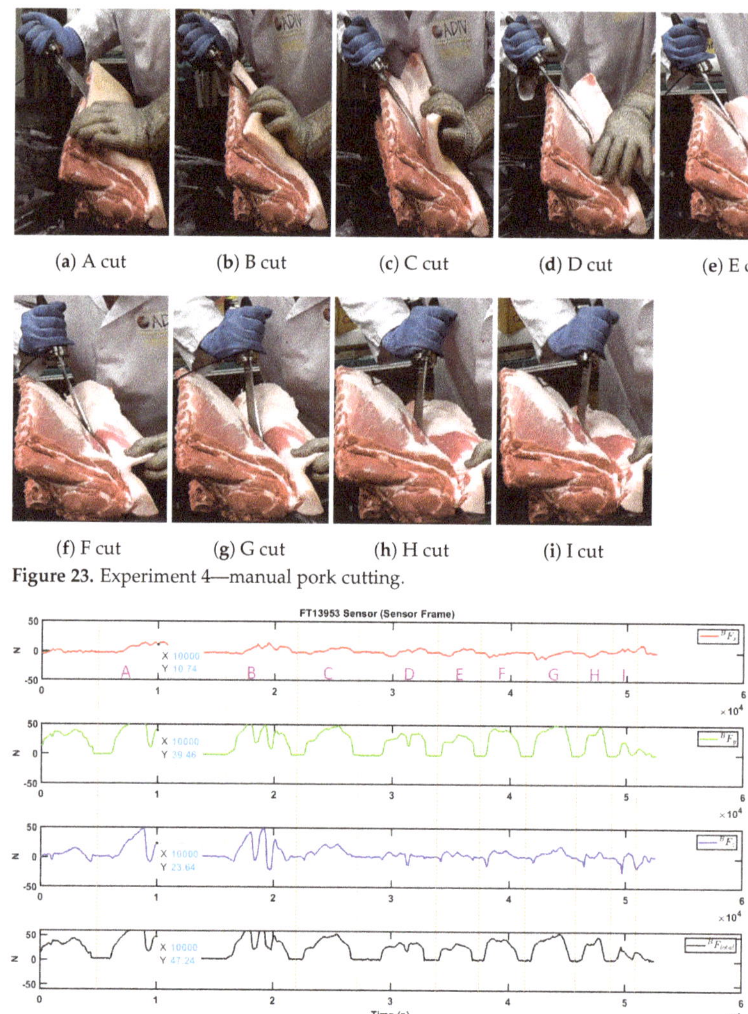

(**a**) A cut (**b**) B cut (**c**) C cut (**d**) D cut (**e**) E cut

(**f**) F cut (**g**) G cut (**h**) H cut (**i**) I cut

Figure 23. Experiment 4—manual pork cutting.

Figure 24. Forces applied by the user in Experiment 4.

Experiment 5—Manual pork cutting

Figure 25 shows the 5 cuts of Experiment 5 and Figure 26 represents the corresponding temporal evolution of the forces applied by the human.

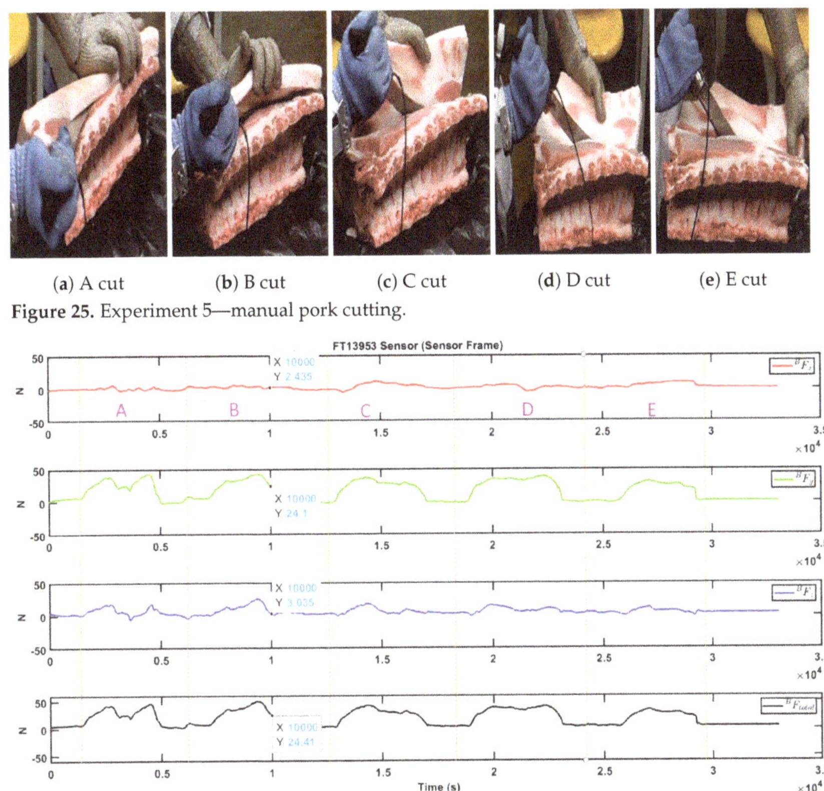

(**a**) A cut   (**b**) B cut   (**c**) C cut   (**d**) D cut   (**e**) E cut

Figure 25. Experiment 5—manual pork cutting.

Figure 26. Forces applied by the user in Experiment 5.

Experiment 6—Manual pork cutting

Finally, Figure 27 shows the 7 cuts of Experiment 6 and Figure 28 represents the corresponding temporal evolution of the forces applied by the human.

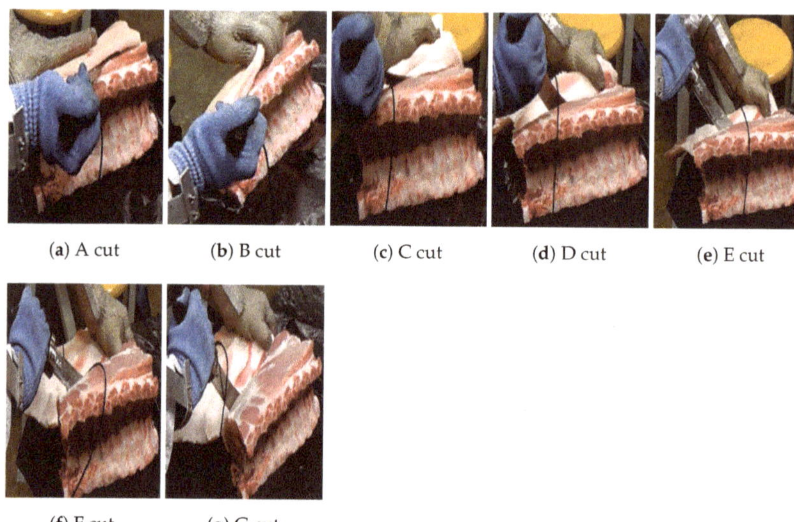

(**a**) A cut   (**b**) B cut   (**c**) C cut   (**d**) D cut   (**e**) E cut

(**f**) F cut   (**g**) G cut

Figure 27. Experiment 6—manual pork cutting.

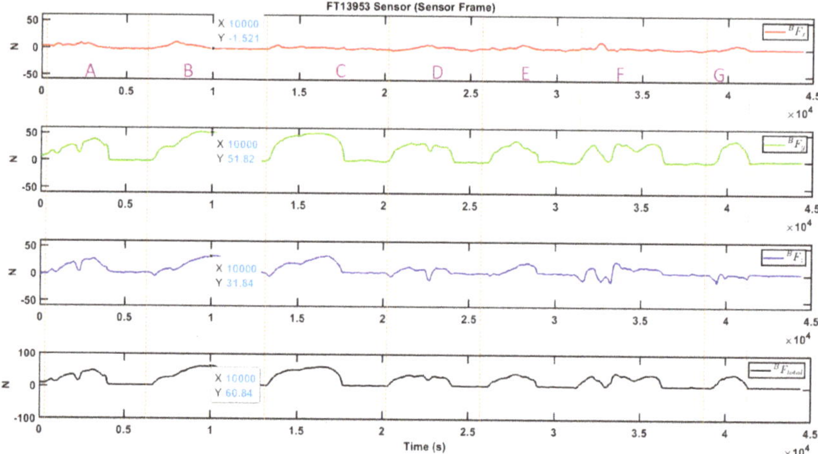

Figure 28. Forces applied by the user in Experiment 6.

Experiment 7—Foam cutting with intent prediction module

For the previous meat cutting experiments with force amplification strategy we had a professional butcher as the user. However, for verifying the intent prediction strategy, we used a foam block and multiple users in order to perform the training of the LSTM network (see Figure 29 for the corresponding experimental setup). This foam reproduces the same type of shearing forces as meat cutting.

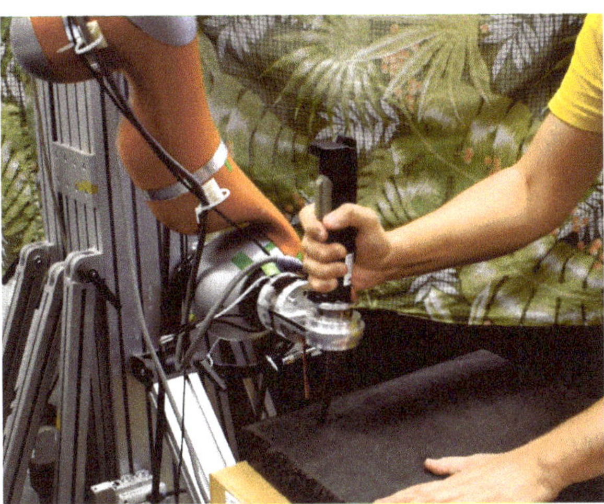

Figure 29. Foam cutting along the global y-direction.

As explained earlier, our intent prediction module uses RNN-LSTM units. For each user, we performed sample trials with foam cutting to collect the training dataset. The LSTM network was trained on this dataset to predict force values, similar to the prediction of force values of the Natural Motion dataset (NM-F) in [23].

For each user we took 90 percent of the sample dataset as the training dataset. The entire architecture consisted of 4 layers- an input layer, an LSTM layer, a fully connected layer and a regression layer. We tested the prediction accuracy with combinations of different hyperparameters such as the number of epochs, sequence length and learning rate. However the results were almost the same i.e., a prediction accuracy of 0.3 (root mean square error).

The number of features was 1 as we had only a single variable—force applied. Similarly as only 1 output was expected, the number of responses was 1. The number of hidden units was taken as 200 and the maximum number of epochs was set to 250. The sequence length for the input layer was 25 time steps (0.2 s). For the output layer, we used Stochastic Gradient Descent algorithm with a learning rate of 0.01. The sigmoid function was used as the activation function for the 3 gates—In, Out and Forget in the LSTM units as it outputs a value between 0 and 1. However for the memory cell, the values should be able to increase or decrease which is not possible with the sigmoid function as the output is always non-negative, hence we used the hyperbolic tangent function (tanh) as the activation function for the memory cell.

Figure 30 shows the plot of cutting forces applied by a user with and without the intent prediction module (for 30 cm cutting of the foam). When the intent prediction module was turned off, we had K_{min} and D_{min} only (i.e. minimum impedance). With the intent prediction module the user applied only 20 percent of the forces as compared to when the module was turned off.

We also compared the force amplification strategy with the intent prediction strategy with 5 users as shown in Figure 31. For each user sample trials were conducted with force amplification strategy for them to decide which amplification factor they are comfortable with. All the users stated that the intent prediction module made the cutting more intuitive than the force amplification strategy.

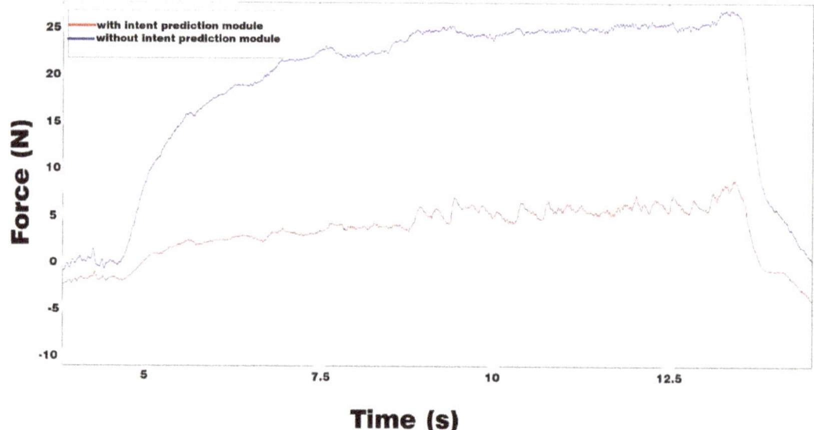

Figure 30. Cutting forces applied by the user using the intent prediction module for 30 cm cutting of the foam. The blue line is the force applied with K_{min} and D_{min} only (i.e. minimum impedance), while the red line is the force applied with K_{min} and D_{min} and the intent prediction module.

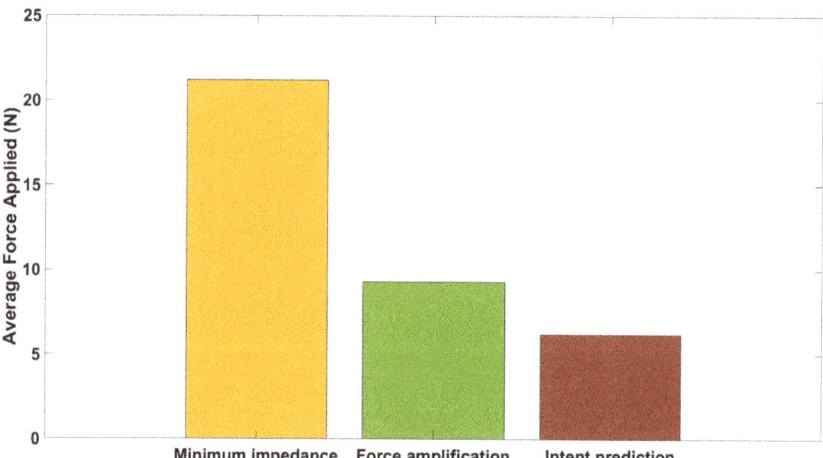

Figure 31. Comparison of different strategies with 5 users.

6. Conclusions and Future Work

In this paper, we demonstrated a proof of concept of two pHRI-based assistive strategies for an industrial meat cutting system. It was determined that sensor B below the joystick gives better reactivity and hence this sensor should be used as the sensor input. From the previous experiments, it is evident that the forces applied by the user are approximately 30% small with the cobot and the force amplification strategy than with the manual meat cutting operation. Blades with lengths 20 cm and 10 cm were tested, and the one with length 10 cm was adjudged to give better cutting performance, because it was stiffer, with less bending and mechanical compliance.

The cartesian impedance controller runs at a frequency of 1000 Hz, as well as the FT sensor. At this frequency the user found it intuitive and useful to operate the tool for the meat cutting task without any issue. However, it is known that the human central nervous system operates at a lower frequency than a robot controller, and if he is coupled to the system via a tool it is possible for the user to 'perceive' a loss of control if the system is too reactive, an observation that capped the upper limit of the amplification in the force amplification strategy.

In the foam cutting experiment, it was shown that the intent prediction strategy was better than the force amplification strategy, especially with regards to intuitiveness. In a pHRI system such as this one, not only should the robot provide assistance as a machine, but also the interface should be intuitive, natural and easy to use.

Our contributions are:

1. We followed a systematic methodology to develop a user friendly pHRI controller system for meat cutting.
2. We developed two assistive strategies: a force amplification strategy and an intent prediction strategy.
3. The developed system allowed the user to move the knife in all 6 degrees of freedom.
4. Using impedance shaping, the impedance values were altered between the cutting and non-cutting re-positioning movement of the knife, which was found as comfortable by the user.

For future work, we would like to compare the force amplification strategy and the intent prediction strategy on meat cutting tasks. Furthermore, in the current experiments we had only one professional butcher and it would be interesting to see what are the experimental results with more professional users. The next step of project Exoscarne would involve the design and development of a specific exoskeleton for meat cutting and transferring the controller that was developed in this work.

Author Contributions: Conceptualization: H.M., J.A.C.R., Y.M.; methodology: H.M., J.A.C.R., L.L., Y.M., M.A.; software, H.M., L.L.; validation and analysis: H.M., M.A.; writing—original draft preparation: H.M.; writing—review and editing: J.A.C.R., L.L., Y.M., M.A.; supervision: J.A.C.R., Y.M.; funding acquisition: J.A.C.R., Y.M., M.A. All authors have read and agreed to the published version of the manuscript.

Funding: This research received funding from the French government research program Investissements d'Avenir through the UMTs ACTIA Mécarnéo-AgRobErgo and the project Exoscarne (Call P3A-ICF2A-2I2A, FranceAgriMer) and from the European Union's Horizon 2020 research and innovation programme under grant agreement n° 869855 (SoftManBot project).

Informed Consent Statement: Informed consent was obtained from all subjects involved in the study.

Conflicts of Interest: The authors declare no conflict of interest.

Abbreviations

The following abbreviations are used in this manuscript:

pHRI	Physical human-robot interaction
MSD	Musculoskeletal disorders
ARMS	A multi arms Robotic system for Muscle Separation
RNN	Recurrent neural networks
LSTM	Long Short Term Memory
ROS	Robot Operating System
FT	Force-Torque
URDF	Unified Robot Description Format
FRI	Fast Research Interface
DH	Denavit–Hartenberg
LWR	Light Weight Robot
KDL	Kinematics and Dynamics Library
IAD	Intelligent Assist Devices

References

1. *Statistiques Accidents du Travail et Maladies Professionnelles*; Carsat Bretagne: Rennes, France, 2020.
2. *Rapport Annuel 2019, L'Assurance Maladie-Risques Professionels, Éléments Statistiques et Financiers*; Caisse nationale de l'Assurance Maladie des travailleurs salariés: Paris, France, 2020.
3. *État de Santé des Salariés de la Filière Viande du Régime Agricole en Bretagne*; Institut de Veille Sanitaire: Saint-Maurice, France, 2019.
4. De Medeiros Esper, I.; From, P.J.; Mason, A. Robotisation and intelligent systems in abattoirs. *Trends Food Sci. Technol.* **2021**, *108*, 214–222. [CrossRef]
5. Long, P.; Khalil, W.; Martinet, P. Modeling control of a meat-cutting robotic cell. In Proceedings of the 2013 16th International Conference on Advanced Robotics (ICAR), Montevideo, Uruguay, 25–29 November 2013; pp. 1–6. [CrossRef]
6. Long, P.; Khalil, W.; Martinet, P. Robotic cutting of soft materials using force control image moments. In Proceedings of the 2014 13th International Conference on Control Automation Robotics Vision (ICARCV), Singapore, 10–12 December 2014; pp. 474–479. [CrossRef]
7. Long, P.; Khalil, W.; Martinet, P. Force/vision control for robotic cutting of soft materials. In Proceedings of the 2014 IEEE/RSJ International Conference on Intelligent Robots and Systems, Chicago, IL, USA, 14–18 September 2014; pp. 4716–4721. [CrossRef]
8. Misimi, E.; Øye, E.R.; Eilertsen, A.; Mathiassen, J.R.; Åsebø, O.B.; Gjerstad, T.; Buljo, J.; Øystein Skotheim. GRIBBOT—Robotic 3D vision-guided harvesting of chicken fillets. *Comput. Electron. Agric.* **2016**, *121*, 84–100. [CrossRef]
9. Wei, G.; Stephan, F.; Aminzadeh, V.; Dai, J.S.; Gogu, G. DEXDEB—Application of DEXtrous Robotic Hands for DEBoning Operation. In *Gearing up and Accelerating Cross-Fertilization Between Academic and Industrial Robotics Research in Europe*; Röhrbein, F., Veiga, G., Natale, C., Eds.; Springer International Publishing: Cham, Switzerland, 2014; pp. 217–235.
10. Alric, M.; Stephan, F.; Sabourin, L.; Subrin, K.; Gogu, G.; Mezouar, Y. Robotic solutions for meat cutting and handling. In *European Workshop on Deformable Object Manipulation*; Innorobo: Lyon, France, 2014.
11. Lawitzky, M.; Mörtl, A.; Hirche, S. Load sharing in human-robot cooperative manipulation. In Proceedings of the 19th International Symposium in Robot and Human Interactive Communication, Viareggio, Italy, 13–15 September 2010; pp. 185–191. [CrossRef]
12. Reinkensmeyer, D.J.; Wolbrecht, E.; Bobrow, J. A computational model of human-robot load sharing during robot-assisted arm movement training after stroke. In Proceedings of the 2007 29th Annual International Conference of the IEEE Engineering in Medicine and Biology Society, Lyon, France, 22–26 August 2007; pp. 4019–4023.
13. Mörtl, A.; Lawitzky, M.; Kucukyilmaz, A.; Sezgin, M.; Basdogan, C.; Hirche, S. The role of roles: Physical cooperation between humans and robots. *Int. J. Robot. Res.* **2012**, *31*, 1656–1674. [CrossRef]
14. Dumora, J.; Geffard, F.; Bidard, C.; Aspragathos, N.A.; Fraisse, P. Robot assistance selection for large object manipulation with a human. In Proceedings of the 2013 IEEE International Conference on Systems, Man, and Cybernetics, Manchester, UK, 13–16 October 2013; pp. 1828–1833.
15. Medina, J.R.; Lawitzky, M.; Molin, A.; Hirche, S. Dynamic strategy selection for physical robotic assistance in partially known tasks. In Proceedings of the 2013 IEEE International Conference on Robotics and Automation, Karlsruhe, Germany, 6–10 May 2013; pp. 1180–1186.
16. Erden, M.S.; Marić, B. Assisting manual welding with robot. *Robot. Comput.-Integr. Manuf.* **2011**, *27*, 818–828. [CrossRef]
17. Colgate, J.E.; Peshkin, M.; Klostermeyer, S.H. Intelligent assist devices in industrial applications: A review. In Proceedings 2003 IEEE/RSJ International Conference on Intelligent Robots and Systems (IROS 2003), Las Vegas, NV, USA, 27–31 October 2003; Volume 3, pp. 2516–2521. [CrossRef]
18. Paxman, J.; Liu, D.; Wu, P.; Dissanayake, G. *Cobotics for Meat Processing: An Investigation into Technologies Enabling Robotic Assistance for Workers in the Meat Processing Industry*; Technical Report; University of Technology: Sydney, Australia, 2006.
19. Campeau-Lecours, A.; Otis, M.J.D.; Gosselin, C. Modeling of physical human–robot interaction: Admittance controllers applied to intelligent assist devices with large payload. *Int. J. Adv. Robot. Syst.* **2016**, *13*. [CrossRef]
20. Lamy, X.; Collédani, F.; Geffard, F.; Measson, Y.; Morel, G. Overcoming human force amplification limitations in comanipulation tasks with industrial robot. In Proceedings of the 2010 8th World Congress on Intelligent Control and Automation, Jinan, China, 7–9 July 2010; pp. 592–598. [CrossRef]
21. Duchaine, V.; Gosselin, C.M. Investigation of human-robot interaction stability using Lyapunov theory. In Proceedings of the 2008 IEEE International Conference on Robotics and Automation, Pasadena, CA, USA, 19–23 May 2008; pp. 2189–2194.
22. Dimeas, F.; Aspragathos, N. Online stability in human-robot cooperation with admittance control. *IEEE Trans. Haptics* **2016**, *9*, 267–278. [CrossRef] [PubMed]
23. Maithani, H.; Ramon, J.A.C.; Mezouar, Y. Predicting Human Intent for Cooperative Physical Human-Robot Interaction Tasks. In Proceedings of the 2019 IEEE 15th International Conference on Control and Automation (ICCA), Edinburgh, UK, 6–9 July 2019; pp. 1523–1528. [CrossRef]
24. Grafakos, S.; Dimeas, F.; Aspragathos, N. Variable admittance control in pHRI using EMG-based arm muscles co-activation. In Proceedings of the 2016 IEEE International Conference on Systems, Man, and Cybernetics (SMC), Budapest, Hungary, 9–12 October 2016; pp. 001900–001905.
25. Hochreiter, S.; Schmidhuber, J. Long short-term memory. *Neural Comput.* **1997**, *9*, 1735–1780. [CrossRef] [PubMed]
26. Albu-Schäffer, A.; Haddadin, S.; Ott, C.; Stemmer, A.; Wimböck, T.; Hirzinger, G. The DLR lightweight robot: Design and control concepts for robots in human environments. *Ind. Robot. Int. J.* **2007**, *34*, 376–385. [CrossRef]

27. Bischoff, R.; Kurth, J.; Schreiber, G.; Koeppe, R.; Albu-Schäffer, A.; Beyer, A.; Eiberger, O.; Haddadin, S.; Stemmer, A.; Grunwald, G.; et al. The KUKA-DLR Lightweight Robot arm-a new reference platform for robotics research and manufacturing. In Proceedings of the ISR 2010 (41st International Symposium on Robotics) and ROBOTIK 2010 (6th German Conference on Robotics), VDE, Munich, Germany, 7–9 June 2010; pp. 1–8.

28. Albu-Schaffer, A.; Hirzinger, G. Cartesian impedance control techniques for torque controlled light-weight robots. In Proceedings of the 2002 IEEE International Conference on Robotics and Automation (Cat. No. 02CH37292), Washington, DC, USA, 11–15 May 2002; Volume 1, pp. 657–663.

29. Schreiber, G.; Stemmer, A.; Bischoff, R. The fast research interface for the kuka lightweight robot. In Proceedings of the IEEE Workshop on Innovative Robot Control Architectures for Demanding (Research) Applications How to Modify and Enhance Commercial Controllers (ICRA 2010), Anchorage, AK, USA, 3–7 May 2010; pp. 15–21.

30. *KUKA LWR Document Manual*; KUKA Roboter GmbH: Augsburg, Germany 2010.

31. Quigley, M.; Conley, K.; Gerkey, B.; Faust, J.; Foote, T.; Leibs, J.; Wheeler, R.; Ng, A.Y. ROS: An open-source Robot Operating System. In *ICRA Workshop on Open Source Software*; IEEE ICRA 2009: Kobe, Japan, 12–17 May 2009; Volume 3, p. 5.

32. Available online: http://docs.ros.org/indigo/api/orocos_kdl/html/ (accessed on 21 April 2021).

33. Corke, P. *Robotics, Vision and Control: Fundamental Algorithms in MATLAB® Second Edition, Completely Revised*; Springer: Berlin/Heidelberg, Germany, 2017; Volume 118.

34. Available online: https://in.mathworks.com/help/robotics/ (accessed on 21 April 2021).

Article

A Hand Motor Skills Rehabilitation for the Injured Implemented on a Social Robot

Francisco Gomez-Donoso *,†, **Felix Escalona** †, **Nadia Nasri** † and **Miguel Cazorla** †

University Institute for Computer Research, University of Alicante, P.O. Box 99, Alicante, Spain;
felix.escalona@ua.es (F.E.); nnasri@dccia.ua.es (N.N.); miguel.cazorla@ua.es (M.C.)
* Correspondence: fgomez@ua.es
† These authors contributed equally to this work.

Featured Application: The system described in this work is intended to be applied to hand motor skill rehabilitation and recovery.

Abstract: In this work, we introduce HaReS, a hand rehabilitation system. Our proposal integrates a series of exercises, jointly developed with a foundation for those with motor and cognitive injuries, that are aimed at improving the skills of patients and the adherence to the rehabilitation plan. Our system takes advantage of a low-cost hand-tracking device to provide a quantitative analysis of the performance of the patient. It also integrates a low-cost surface electromyography (sEMG) sensor in order to provide insight about which muscles are being activated while completing the exercises. It is also modular and can be deployed on a social robot. We tested our proposal in two different facilities for rehabilitation with high success. The therapists and patients felt more motivation while using HaReS, which improved the adherence to the rehabilitation plan. In addition, the therapists were able to provide services to more patients than when they used their traditional methodology.

Keywords: hand motor rehabilitation; sEMG; hand pose; social robot

Citation: Gomez-Donoso, F.; Escalona, F.; Nasri, N.; Cazorla, M. A Hand Motor Skills Rehabilitation for the Injured Implemented on a Social Robot. *Appl. Sci.* **2021**, *11*, 2943. https://doi.org/10.3390/app11072943

Academic Editors: Ehud Ahissar and Juan Antonio Corrales Ramón

Received: 19 February 2021
Accepted: 24 March 2021
Published: 25 March 2021

Publisher's Note: MDPI stays neutral with regard to jurisdictional claims in published maps and institutional affiliations.

1. Introduction

Rehabilitation of the brain- and motor-injured is an important task. These kinds of challenged individuals have reduced motion in their muscles as a result of an accident, acquired diseases, or birth conditions. However, they can improve their motor skills by following a rehabilitation plan. Nonetheless, foundations such as ADACEA, which is a Spanish-based organization for the brain- and motor-injured, are usually under founded and short staffed, and sometimes they cannot provide the required rehabilitation services. In addition, there is a lack of an established protocol to quantitively evaluate a patient's performance on rehabilitation exercises.

So far, the therapists of ADACEA carry out exercises with patients one by one, and do not have any quantitative method to evaluate the patients' performance. The patients have the rehabilitation sessions scheduled, but since these kinds of foundations are usually short staffed, as mentioned before, the patients do not execute the amount of rehabilitation sessions they should. In addition, the evaluation is purely qualitative. In this sense, the therapists take notes about the performance and significant events such as pain or unusually poor execution of the exercises based on the therapists' experience. These eventualities negatively impact the way they follow the rehabilitation process and what they take away from it.

Having this need in mind, we developed HaReS, a hand rehabilitation system, which is a system for motor rehabilitation. By using HaReS, patients can perform rehabilitation exercises on their own, without requiring the presence of a therapist. Furthermore, HaReS automatically grades the performance of users, so therapists have a quantitative measure-

89

ment of patients' performance. This way, the therapists can provide an optimized and enhanced service to more patients.

HaReS is composed of a surface electromyography (sEMG) sensor and a hand-tracking device that allows to record and provide quantitative measurements and feedback of the rehabilitation sessions that a patient is performing. The exercises are set up by a therapist for each user. In addition, HaReS can take advantage of a social robot to show the exercises and to interact with patients. Thus, enhancing their adherence and engaging them. The system is modular, so it is able to work without the sensors or the robot. Apart from the robot, the rest of the system is fairly low cost. It is important to note that it is executed on a desktop or laptop computer, so all components are connected to it. The hand-tracking device, the sEMG sensor, and the social robot are all connected to the computer, which is in charge of running HaReS. In this sense, the robot is not executing HaReS, nor are the sensors connected to it. The role of the robot within the HaReS framework is to display the system in its built-in screen and interact with the patient using gestures, lights, and its speech capabilities.

The main contributions of this paper are the following:

- A hand rehabilitation system that integrates a series of exercises for the motor- and brain-injured;
- Use of low-cost sensors for hand tracking and muscle signal monitoring;
- An automatic and quantitative evaluation of the exercises to be analyzed by the therapists;
- Optional use of a social robot to improve patient adherence to rehabilitation

The rest of the paper is structured as follows. First, some related work to this matter are presented in Section 2. Then, Section 3 describes the proposal in detail. Next, the evaluation we performed to validate the system is given in Section 4. Finally, Section 5 states the conclusion of this work and future research directions.

2. Related Works

As stated in [1], conventional exercise programs follow the Bobath [2,3] or Brunnstrom [4] concepts, as well as the proprioceptive neuromuscular facilitation (PNF) principles [5]. On the one hand, Bobath's method emphasizes the reduction of enhanced muscle tone before facilitating active movements by means of cutaneous and proprioceptive stimuli applied to the region of the target muscles. On the other hand, Brunnstrom's approach and PNF use maximal innervation of intact or less-paretic muscle groups to produce irradiation effects in more severely paretic synergistic muscle groups. Several studies have been carried out with some of these techniques [6–10]. However, none of these studies presented a suitable control group, so the benefit was difficult to extrapolate from spontaneous recovery. Additionally, they presented a lack of exercise strategies for the hand muscles.

In [1], the authors investigated the effect of a specific training program focused on the performance of the basic movement parameters of the hand. They paid special attention to the identical repetition of movements, whose long-term benefits have been demonstrated in other studies with animals [11,12]. The patient scores for the rehabilitation exercises were calculated according to the Rivermead Motor Assessment [13], a widely used technique to measure motor ability in stroke patients, and 24 out of 27 patients showed an important improvement according to this metric. In this study, they introduced control groups to ensure the impact of the specific repetitive motor training in motion improvement.

In order to evaluate muscles' progress during the rehabilitation process, some rehabilitation exercises take advantage of surface electromyography (sEMG) electrodes [14–17]. These wearable electrodes enable the professional in charge to analyze the muscular activities and muscular strength of a patient during each exercise [18,19]. In some rehabilitation methods, with aim of having an interactive system and motivating patients with visual and audio feedback, during the process, sEMG analysis is incorporated into a 3D game [20,21], virtual reality (VR) [22], and augmented reality (AR) [23]. In addition, there are algorithms that take sEMG signals as the input in order to compute [24] different low-level movements of the muscles. Specifically, this paper introduces a method to perform movement identification of the upper limb: Abduction, adduction, flexion, extension, and abduction followed by arm to the front.

In recent years, researchers have focused on robot-assisted neurorehabilitation in order to improve the performance of rehabilitation exercises. According to [25,26], two main approaches have been used to design hand rehabilitation devices: End-effector and exoskeleton robots.

End-effector-based robots are usually employed to simulate activities of daily living (ADLs), to train the hand and, eventually, the wrist function by only interacting with the distal segments of the fingers, that is, the fingertips or sometimes the middle phalanxes. The rest of the arm is not controlled by the robot, which may result in patients using undesired compensatory strategies, so it is common to provide a weight support in order to reduce the muscle fatigue produced by the distal limb [27–33]. These types of robots are prepared for hands of different sizes, so they can be easily adjusted for different patients.

On the contrary, an exoskeleton consists of a mechanical structure that is mounted in parallel with the limb of the user in order to provide assistance, so each degree of freedom is locally aligned with the corresponding human joint [16,17,34–40]. The majority of hand exoskeletons aim to restore grasping function by helping patients open the hand or provide force augmentation to hold objects. They do not limit the natural joint movements of patients, but their design is far more complex and the adaptation process for new users is more difficult.

Several clinical studies have studied the efficacy of these types of assisted rehabilitation strategies and have concluded that they are effective in reducing the motor impairments of stroke victims, and that they improve the ability to perform activities of daily living [25,41].

Most of these rehabilitation exercises require the presence of professional therapists that help and supervise the results of this process. However, the lack of medical resources, the difficulties in visiting these professionals daily, and the challenges to transfer this clinic technology to a home environment make the rehabilitation process much harder for patients when they are at home.

Recent works have focused on at-home rehabilitation exercises in order to provide these tools outside of a clinic environment. In [42], the authors developed a multisensor system for rehabilitation and interaction with people with motor and cognitive disabilities. In the case of [43], PHAROS was proposed, an interactive robot system that recommends and monitors physical exercises at home, designed for the rehabilitation of chronic diseases. More recently, [44] proposed an augmented reality platform to engage and supervise rehabilitation sessions at home using low-cost sensors. This platform also stores a user's statistics and allows therapists to tailor the exercise programs according to their performance. The results presented in these research works suggest that these platforms improve the results and adhesion to rehabilitation exercises.

However, there is a lack of this type of research specialized in hand motor rehabilitation at home that take into account the particularities and needs of its exercises and evaluations.

Hand pose estimation methods are a really interesting topic, and thus, there are several different state-of-the-art approaches. For instance, there are pure vision-based methods [45,46] that take as input images of hands and compute the hand pose. Nonetheless, these systems lack enough accuracy and are very computationally demanding, requiring a powerful graphics processing unit. There are also algorithms that process electromyography (EMG) signals [47] to predict static hand poses. These kind of methods are also unsuitable because they do not provide fine hand tracking, but rather, a range of fixed poses. Hand tracking can also be achieved by inertial measure unit sensors [48], but it is also limited to a range of specific hand movements. The methods reviewed so far rely on computational models to perform predictions. Nonetheless, there are also hardware-based methods to perform hand tracking. For instance, motion capture systems [49,50] could be used, but they are expensive and intrusive, as they must be worn by the users. After carefully considering the different approaches to perform hand tracking, we decided to include the Leap Motion controller [51] in HaReS. This device is low cost and is not intrusive, as it is placed in the desk and the user does not need to wear it or any specific marker. In addition, its error ranges from 0.2 to 1.2 mm [52], reportedly being accurate and robust enough for a range of different applications, including robot manipulation [53] and dataset creation [54].

3. Architecture of HaReS, the Hand Rehabilitation System

As introduced before, HaReS is a system for hand motor rehabilitation. It is composed of pieces of hardware and software. Regarding the hardware, HaReS uses an sEMG sensor in order to track the activation of the muscles during the rehabilitation session. Then, a hand-tracking device is also used to provide the user's hand poses. The rehabilitation exercises are set up on a software app, which record sthe data of the sessions, including sEMG profiles and hand poses. The records are available for the therapists to review remotely and/or in the future. In addition, some quantitative measurements are given by the system as a potential indicator of the user's performance.

These functionalities enable a better evaluation of the improvement the user experiences by completing the rehabilitation exercises. In addition, the system is modular, so any of the pieces can be removed at any time. Regardless, it is worth mentioning that the devices we used to create HaReS are fairly low cost. Finally, HaReS can be deployed on a social robot for further improving patients' experience and motivation. A diagram of the complete architecture of HaReS is shown in Figure 1.

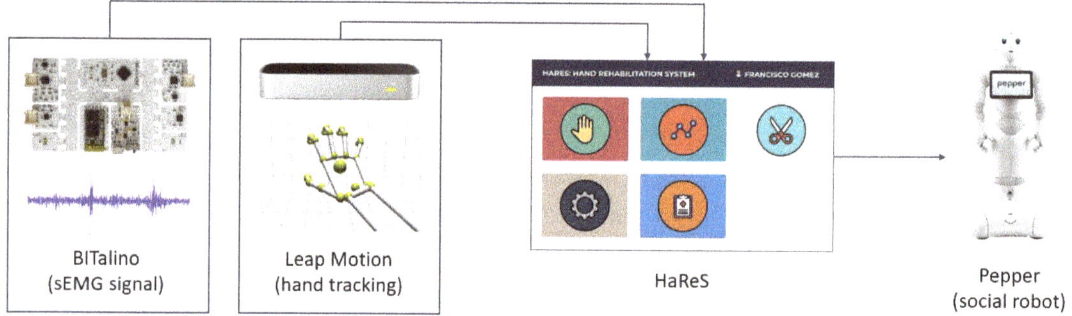

Figure 1. Architecture of HaReS, a hand rehabilitation system, featuring surface electromyography (sEMG) and hand tracking low-cost sensors.

At this point, it is worth noting that the use of a robot is completely optional, as we intended HaReS to be low cost and fully modular. The system can be deployed on a desktop or laptop computer, with no robot involved. We would also like to state that the chosen sensors, such as Leap Motion, are accurate, whist keeping cost at bay. For instance, hand tracking based on cameras is possible [45,46], but lacks enough accuracy and requires much more computation power.

The components of the system are detailed in the following subsections.

3.1. sEMG Sensor

Electromyography (EMG) is a technique used in clinical processes to study the electrical activity produced by skeletal muscles. The signals are captured by a series of electrodes that are placed in the muscle's belly. Surface EMG (sEMG) is a non-invasive type of EMG, and the electrodes should be positioned on the skin surface over the muscles to record the signals. Although this superficial type of EMG is less accurate than intramuscular EMG (invasive type), it is a better fit for our purpose. sEMG electrodes are user-friendly and they are accurate enough to capture muscle activity during exercise.

The sensor of choice for HaReS esd based on the BITalino project. This device aims to provide a framework for science education, research, and prototyping. It includes several sensors and actuators such as electrocardiogram (ECG), electrodermal activity (EDA), electroencephalogram (EEG) and, of course, electromyogram (EMG). The sEMG sensor of the BITalino is bipolar, which means that it has two electrodes for sensing the electrical activity of the muscle and one as a reference. The measuring range is ± 1.64 mV and the sampling rate is up to 4000 Hz.

As suggested in [55], the optimal sample rate for EMG pattern recognition is between 400 and 500 Hz, so we used 500 Hz as the sampling rate for the data acquisition process that involved the BITalino sensor. To study the finger and wrist movements, the electrodes were placed on the flexor carpi radialis. In addition to the electrodes' placement, we were able to receive data from contraction of the flexor digitorum, and for the reference, we placed an electrode on the elbow bone.

This sensor is used by HaReS to record the sEMG activity of a desired muscle. Given these data, along with the synchronized motion data captured by a hand-tracking sensor, an experienced therapist could detect if the patient is triggering the correct muscles during the performance of a certain task, thus having more insight about whether the rehabilitation program is improving the patient's skills. Figure 2 shows some sample hand poses with the corresponding sEMG data.

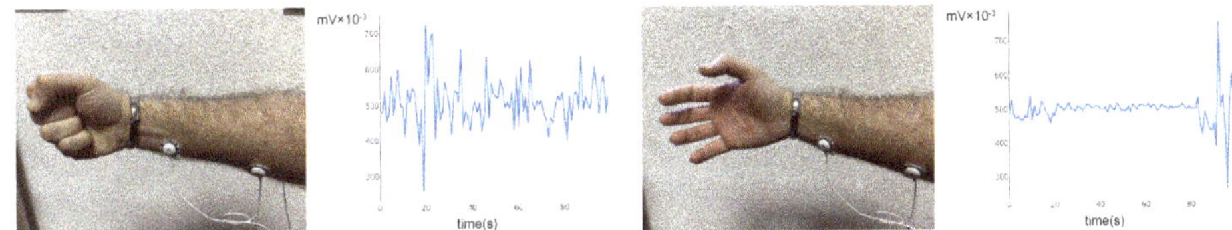

Figure 2. Two samples of hand poses and the corresponding sEMG profile, as provided by the BITalino controller. The X-axis shows the time, measured in seconds, and the Y-axis shows the muscle activity, measured in mV $\times 10^{-3}$.

3.2. Hand-Tracking Sensor

Another important feature of HaReS is the ability to review the user's hand poses while performing the rehabilitation exercises. There are different approaches to tackle this problem, but we aimed for two important features. First, it had to be low cost so that the system could be implemented by anyone. Next, it also had to be non-invasive so that it would not interfere with the already limited motor skills of the patients.

Thus, the sensor of choice for hand tracking was the Leap Motion controller. This device yields an infrared-based stereo setup, which allows to compute the position and orientation of the bones in a hand, namely, the user's hand pose. Leap Motion sends data at 115 Hz and is precise enough to be considered ground truth [54]. Some sampless ass provided by this sensor can be seen in Figure 3.

This functionality is used by HaReS to record the user's hand poses during the exercises. It provides therapists with more data to improve the evaluation of the rehabilitation program and to know whether patients are performing the exercises correctly.

3.3. Social Robot

The social robot within the HaReS framework is used for displaying information and interacting with the patient. The inclusion of this robot is merely intended to improve the motivation of patients and to enhance their experience while using HaReS.

In this regard, the social robot of choice was the Pepper robot. Pepper is an humanoid-like robot that focuses on interaction. It yields LEDs on its face and has arms that enable it to show emotions and engage with user interaction. It also has a touchscreen built-in tablet that allows to display HaReS.

We used the Pepper robot within the HaReS framework to run the software on its tablet, to cheer up and encourage patients, and to provide feedback using its arms, face, and speech capabilities.

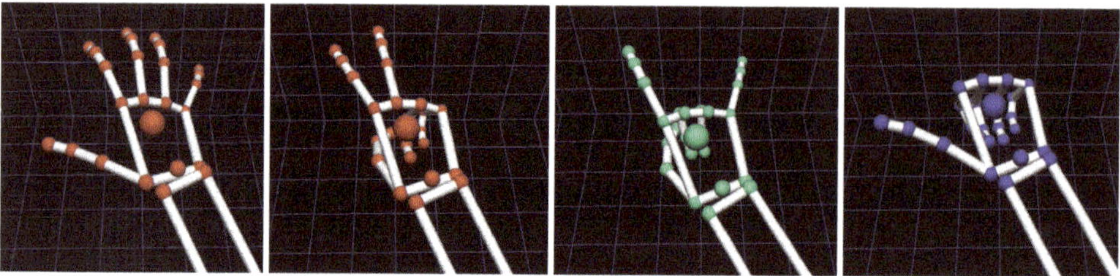

Figure 3. Some samples of hand poses, as provided by the Leap Motion controller.

3.4. Rehabilitation Exercises

As stated before, HaReS has a range of rehabilitation exercises. These exercises were jointly designed with therapists of ADACEA (Alicante, Spain), which is a foundation for the rehabilitation of acquired brain injury-affected individuals.

So far, for the motor rehabilitation of the hands, they divide the exercises in two categories. The first category focuses on strengthening the muscles on the hand and on low-level and fine motion. This part of the rehabilitation ensures that patients are able to perform the correct movements. The second part focuses on functional skills that ensure patients learn how to correctly interact with everyday objects. The tasks designed by the therapists of ADACEA and implemented in HaReS cover the first part entirely and provide entry-level exercises to cover the functional part. It is important to note that therapists must set up the exercise schedule and the settings for each exercise for each user.

The rehabilitation exercises integrated in HaReS are detailed in the following subsections.

3.4.1. Copy Pose

In this exercise, the patient must copy a static hand pose for a configurable amount of time. The goal hand poses are recorded using the Leap Motion device by the therapists. The same device is used for processing the patient's hand pose whilst performing this exercise. A tolerance threshold is applied in order to consider that a pose is correct.

As explained before, the Leap Motion controller provides the tridimensional position of each joint in a hand, taking advantage of this to measure the users' performance. In this case, we computed the mean euclidean distance to each fingertip from the user's pose to the goal pose. In order to make the system tolerant to different users, the hands were normalized so that they all had the same size. This is also adequate for measurements, as it enables a global framework of comparison that is not user-dependent.

An image depicting a subject executing this exercise can be seen on the leftmost image of Figure 4.

3.4.2. Copy Pose Dynamic

This exercise is similar to the one described in Section 3.4.1. However, in this case, the poses are dynamic, namely, the user must copy a specific movement instead of a fixed pose. The dynamic hand poses the patient must replicate are prerecorded by the very same hand-tracking device that is used in the HaReS system, namely, the Leap Motion controller. The therapists can thus add new dynamic poses to the task with ease so that the patients can replicate them.

To quantitatively rank the similarity between the goal movement and the user movement, we computed dynamic time warping (DTW) [56]. Applied to this scope, DTW measures the similarity between the trajectories of two 3D points that are shifted on the time dimension, namely, it isolates the time factor measuring only the similarity of two trajectories. As the original implementation is very computationally demanding, we used an approximation of the algorithm, which is much faster, as described in [57].

We applied DTW to the trajectories of the fingertips between the goal movement and the user movement and averaged them. An exercise is considered correct if this value is under a configurable tolerance threshold.

The rightmost image of Figure 4 depicts a subject carrying out one of these exercises.

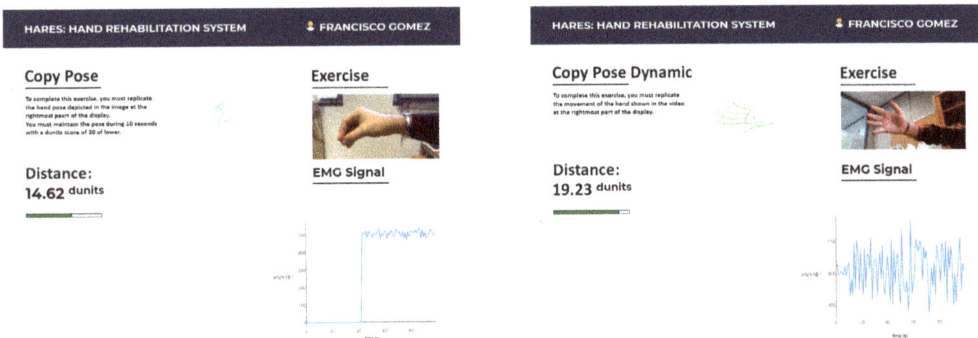

Figure 4. HaReS screenshots that correspond to a copy pose exercise (leftmost) and a copy pose dynamic exercise.

3.4.3. Follow the Path

This exercise consists of following a specific path with the index fingertip of the hand, namely, the user must place the finger on an array of subgoals arranged on the screen sequentially. As providing visual feedback within the display of a tridimensional position is ambiguous, the paths are simplified to be 2D. Thus, the paths are defined in a plane that is perpendicular to the ground.

In this case, the time the user takes to reach all of the subgoals is measured. As explained before, the euclidean distance is computed between the subgoal and the 2D fingertip position, and if the distance is lower than a threshold, the user is considered to have reached that subgoal.

The leftmost image of Figure 5 depicts a subject carrying out one of these exercises.

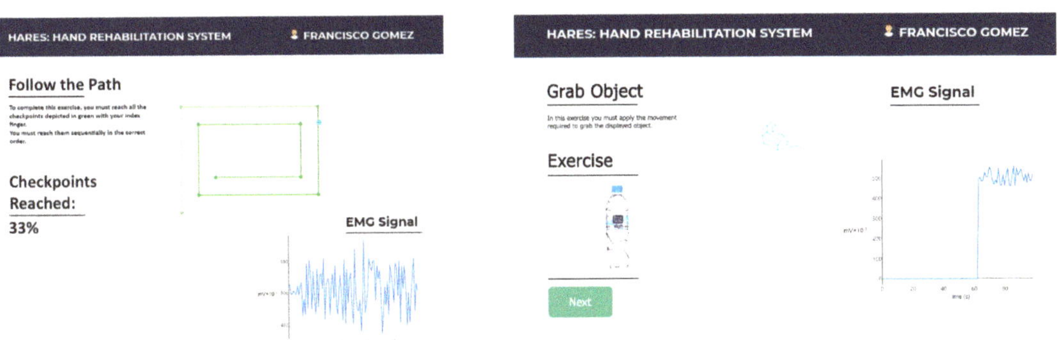

Figure 5. HaReS screenshots that correspond to a follow the path exercise (leftmost) and a grab object exercise.

3.4.4. Grab Object

Unlike the exercises described in the sections above, which are focused on strengthening the muscles on the hand and on low-level and fine motion, this exercise also introduces a functional factor as well. In this exercise, an object is displayed to the patient, who performs the movement required to grab the object. As there is a range of different ways to grab a certain object, no quantitative measure is provided for these exercises.

As the sessions are recorded, they provide the therapists with evidence about the patient's skills, and whether their grabbing methods harm other muscles due to unnatural movements induced by the condition of the patient.

A subject carrying out one of these exercises can be seen in the rightmost image of Figure 5.

3.4.5. Free Style

This task does not have a fixed aim, but instead records the patients' hand movements for the therapists so that they can review them later. This option enables the therapist to ask for certain exercises that are not considered by HaReS and still provide valuable insight about the user rehabilitation program.

4. Evaluation

In order to evaluate the value of HaReS, we conducted complete pilot tests in two different facilities for motor and cognitive rehabilitation. One was carried out in the ADACEA foundation, as mentioned before (five patients with a range of different levels of hand motor limitation and of different ages, and two therapists), and the second one was performed in a local nursing home (five patients with a range of different levels of hand motor limitation, aged 72–79, and two therapists). The complete pilot test included two sessions of two hours with the robot in each facility at the beginning and at the end of the pilot test. During the remaining time, they used HaReS without the social robot, namely, deployed on a desktop computer. In total, the pilot test was carried out within a timeframe of four months.

We divided the evaluation section into the two following subsections. First, we conducted a survey for both patients and therapists, the results of which are shown in Section 4.1, in order to gain insight into the overall benefits of HaReS and the facets that are hard to measure, such as the motivation factor. The next part of the evaluation showed more qualitative aspects of the pilot experiment, such as the evolution of the scores per activity included in HaReS. This matter is shown in Section 4.2.

4.1. Qualitative Evaluation

As mentioned before, surveys were handed to the hosts of the experiments in order to qualitatively evaluate HaReS as a whole system. The survey covered a range of different aspects, from the usability to the impact in the rehabilitation. The results are shown in Table 1. The score was averaged among the answers of both organizations and were in the range of 1–5, 1 being totally disagree and 5 being fully agree.

Table 1. Questions and corresponding scores of the survey conducted in two different facilities for motor and cognitive rehabilitation.

ID	Question	Score
1	The exercises implemented in HaReS overall helped to improve the motor skills of the patients	4
2	The exercises implemented in HaReS helped to improve the adherence at home to the rehabilitation plan	5
3	The exercises implemented in HaReS effectively cover a common rehabilitation plan	3
4	The hand poses provided by the Leap Motion controller are accurate enough	4.5
5	The HaReS interface is comprehensive and easy to use	4
6	HaReS is a motivating factor for the therapists	5
7	HaReS is a motivating factor for the patients	5
8	The use of a social robot is critical in terms of motivation for the therapists	2
9	The use of a social robot is critical in terms of motivation for the patients	4
10	HaReS helped the therapists to provide a better service	4
11	HaReS helped the therapists to provide service to more patients	3

Regarding the results of the survey, the therapists and patients agreed that HaReS helped to improve the motivation of the two groups. Therapists remarked that HaReS indeed helped to improve the motor skills of the patients at the same level of the traditional rehabilitation methodology, and also stated that the proposal helped to provide services to more patients at the same time. For the patients, the use of a social robot was a motivation factor that improved the user experience and engagement. Nonetheless, the therapists stated that the robot had no effect in their motivation, nor on the results of the rehabilitation plan. Regarding the accuracy of the hand poses, the therapists stated that it was acceptable, and that it seldom lost track of the hands or depicted incorrect poses.

In addition to the survey, the therapists also offered valuable feedback. The comments included that HaReS could benefit from more exercises. They also suggested to somehow involve gamification to further improve the adherence to the rehabilitation plan and to enhance the appeal of HaReS. Finally, they proposed the use of a virtual reality setup that provides a better immersion of the patients in order to further improve their experience and the efficiency of the functional exercises.

As a conclusion, we can mention that HaReS indeed helped in the motivation of both therapists and patients, and that HaReS helped to optimize the time that the therapists used to provide services to more patients at the same time.

4.2. Quantitative Evaluation

Here, we show the quantitative results obtained by the patients during the four-month timeframe in which the pilot experiment was conducted. Specifically, we show the quantitative measures of each activity included in HaReS. First, as explained before, the copy pose activity measures the mean euclidean distance from the user's fingertip to the goal. Then, the copy pose dynamic activity measures the mean DTW from the user's fingertip to the goal. Finally, the follow the path activity measures the time the user takes to reach all of the goals. The results are shown in Figures 6–8.

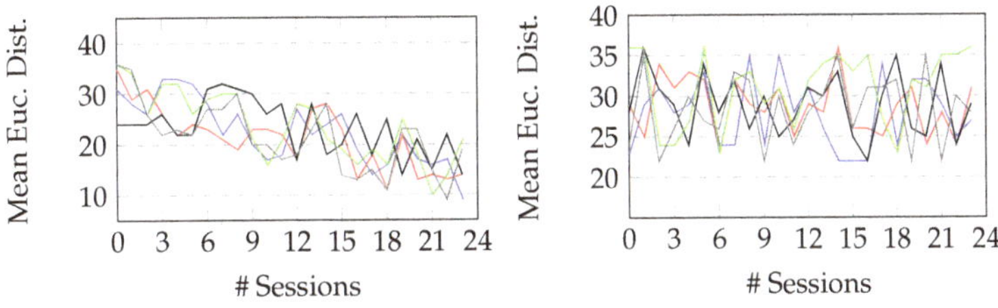

Figure 6. Evolution of the copy pose activity at ADACEA (**left**) and at the nursing home (**right**).

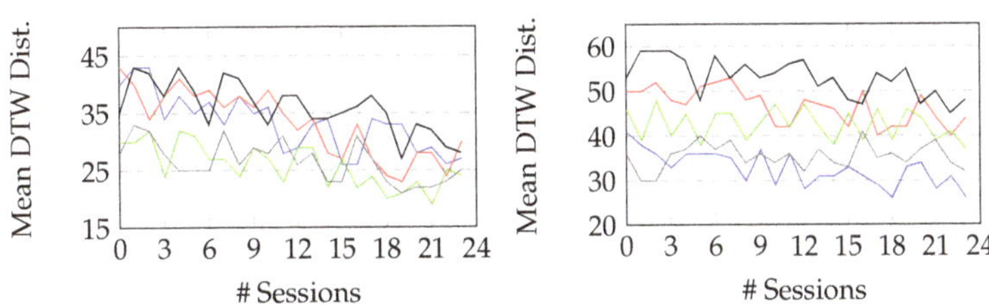

Figure 7. Evolution of the copy pose dynamic activity at ADACEA (**left**) and at the nursing home (**right**).

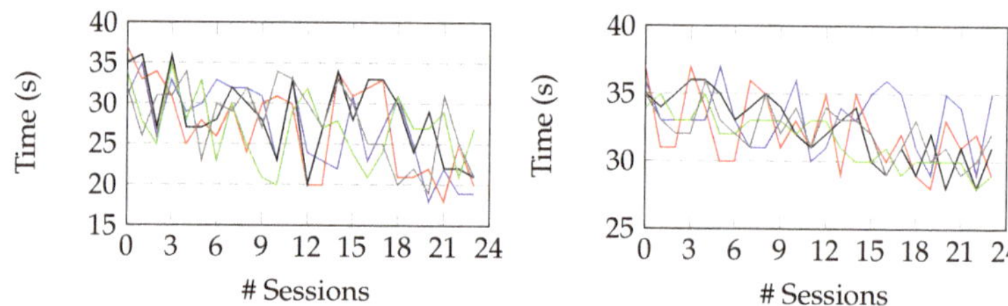

Figure 8. Evolution of the follow the path activity at ADACEA (**left**) and at the nursing home (**right**).

As per the experiments shown, the patients of the ADACEA foundation experienced a significant improvement over the timeframe in which they used HaReS in all tasks. The patients of ADACEA are people that have suffered an accident or a sudden condition. Thus, their hand motor limitation was acquired and, in most cases, reversible. This means that they were expected to improve over the three months in which the experiment was conducted. Nonetheless, the results for the patients of the nursing home were different. They barely experienced any improvement. This is likely due to all of them being elderly people, with motor limitations due to their age. Note that the expected goal of HaReS in this case is to stop further degradation of their muscles.

5. Conclusions and Future Work

In this paper, we introduced HaReS, a hand skill rehabilitation software for the motor- and brain-injured. HaReS takes advantage of low-cost sensors, such as a hand-tracking device and sEMG sensors, to help quantitatively and automatically evaluate the performance of a rehabilitation plan, and helps to follow the evolution of a patient. As there is no need for a therapist to be present during the rehabilitation sessions, it helps to provide services to more patients.

The evaluation we carried out in two different rehabilitation facilities indicated that HaReS effectively helped both therapists and patients with the rehabilitation plan in terms of both adhesion and improvement of the motor skills. In addition, they both found HaReS a motivating factor that engaged users in the rehabilitation exercises.

As for future work, we plan to involve a full machine learning method for estimating users' hand poses that enables to provide accurate prediction whilst keeping the computation cost at bay. In addition, as the therapists stated, it is important that the rehabilitation exercises have a goal, such as grasping an object or completing a puzzle, rather than aimlessly repeating a movement. In this regard, we plan to implement the idea developed here using a virtual reality environment so that patients can engage cognitively even further. In addition, we plan to use the data gathered by HaReS to train machine learning methods that can automatically predict whether the user is improving or not by reviewing their history, or to state whether a user is prone to suffer any hand motor or cognitive disease such as Parkinson's. We plan to also involve computational models [58–60] to produce EMG signals that would enable to provide an automatic analysis to assist therapists to understand the sEMG signals provided by HaReS.

Finally, we want to also remark that due to the COVID-19 constraints, we were unable to access more patients or more rehabilitation facilities. We plan to carry out a proper set of experiments once we are able to freely access these people again.

Author Contributions: Conceptualization, methodology, writing–review and editing, supervision, project administration, and funding acquisition, M.C.; software, validation, formal analysis, resources, visualization, and writing–original draft preparation, F.E.; conceptualization, software, validation, formal analysis, and data curation, N.N.; conceptualization, writing–original draft preparation, software, validation, formal analysis, data curation, and visualization, F.G.-D. All authors read and agreed to the published version of the manuscript.

Funding: This work was funded by a Spanish Government PID2019-104818RB-I00 grant, supported by Feder funds. It was also supported by Spanish grants for PhD studies ACIF/2017/243 and FPU16/00887.

Informed Consent Statement: Informed consent was obtained from all subjects involved in the study.

Acknowledgments: The authors would like to thank NVIDIA (Santa Clara, CA, USA) for the generous donation of a Titan Xp and a Quadro P6000.

Conflicts of Interest: The authors declare no conflict of interest.

References

1. Bütefisch, C.; Hummelsheim, H.; Denzler, P.; Mauritz, K.H. Repetitive training of isolated movements improves the outcome of motor rehabilitation of the centrally paretic hand. *J. Neurol. Sci.* **1995**, *130*, 59–68. [CrossRef]
2. Bobath, B. *Abnormal Postural Reflex Activity Caused by Brain Lesions*; Aspen Publishers: Alexandria, VA, USA, 1985.
3. Bobath, B. *Adult Hemiplegia: Evaluation and Treatment*; William Heinnemann: London, UK, 1978.
4. Brunnstrom, S. *Movement Therapy in Hemiplegia: Neurophysiological Approach*; Harper & Row: New York, NY, USA, 1970; pp. 113–122.
5. Voss, D.E. Proprioceptive neuromuscular facilitation. *Am. J. Phys. Med. Rehabil.* **1967**, *46*, 838–898.
6. Stern, P.H.; McDowell, F.; Miller, J.M.; Robinson, M. Effects of facilitation exercise techniques in stroke rehabilitation. *Arch. Phys. Med. Rehabil.* **1970**, *51*, 526–531. [PubMed]
7. Logigian, M.K.; Samuels, M.; Falconer, J.; Zagar, R. Clinical exercise trial for stroke patients. *Arch. Phys. Med. Rehabil.* **1983**, *64*, 364–367.
8. Dickstein, R.; Hocherman, S.; Pillar, T.; Shaham, R. Stroke rehabilitation: Three exercise therapy approaches. *Phys. Ther.* **1986**, *66*, 1233–1238. [CrossRef] [PubMed]
9. Basmajian, J.; Gowland, C.; Finlayson, M.; Hall, A.; Swanson, L.; Stratford, P.; Trotter, J.; Brandstater, M. Stroke treatment: Comparison of integrated behavioral-physical therapy vs traditional physical therapy programs. *Arch. Phys. Med. Rehabil.* **1987**, *68*, 267–272. [PubMed]
10. Wagenaar, R.; Meijer, O.; Van Wieringen, P.; Kuik, D.; Hazenberg, G.; Lindeboom, J.; Wichers, F.; Rijswijk, H. The functional recovery of stroke: A comparison between neuro-developmental treatment and the Brunnstrom method. *Scand. J. Rehabil. Med.* **1990**, *22*, 1–8.
11. Asanuma, H.; Keller, A. Neuronal mechanisms of motor learning in mammals. *Neurorep. Int. J. Rapid Commun. Res. Neurosci.* **1991**, *2*, 217–224.
12. Asanuma, H.; Keller, A. Neurobiological basis of motor learning and memory. *Concepts Neurosci.* **1991**, *2*, 1–30.
13. Lincoln, N. *Assessment of Motor Function in Stroke Patients*; Physiotherapy, Elsevier: Amsterdam, The Netherlands, 1979.
14. John, J.; Wild, J.; Franklin, T.D.; Woods, G.W. Patellar pain and quadriceps rehabilitation: An EMG study. *Am. J. Sport. Med.* **1982**, *10*, 12–15.
15. Milner-Brown, H.; Stein, R. The relation between the surface electromyogram and muscular force. *J. Physiol.* **1975**, *246*, 549–569. [CrossRef]
16. Hu, X.; Tong, K.; Wei, X.; Rong, W.; Susanto, E.; Ho, S. The effects of post-stroke upper-limb training with an electromyography (EMG)-driven hand robot. *J. Electromyogr. Kinesiol.* **2013**, *23*, 1065–1074. [CrossRef] [PubMed]
17. Leonardis, D.; Barsotti, M.; Loconsole, C.; Solazzi, M.; Troncossi, M.; Mazzotti, C.; Castelli, V.P.; Procopio, C.; Lamola, G.; Chisari, C.; et al. An EMG-controlled robotic hand exoskeleton for bilateral rehabilitation. *IEEE Trans. Haptics* **2015**, *8*, 140–151. [CrossRef] [PubMed]
18. Liu, L.; Chen, X.; Lu, Z.; Cao, S.; Wu, D.; Zhang, X. Development of an EMG-ACC-Based Upper Limb Rehabilitation Training System. *IEEE Trans. Neural Syst. Rehabil. Eng.* **2017**, *25*, 244–253. [CrossRef]
19. Ganesan, Y.; Gobee, S.; Durairajah, V. Development of an Upper Limb Exoskeleton for Rehabilitation with Feedback from EMG and IMU Sensor. *Procedia Comput. Sci.* **2015**, *76*, 53–59. [CrossRef]
20. Nasri, N.; Orts-Escolano, S.; Cazorla, M. An sEMG-Controlled 3D Game for Rehabilitation Therapies: Real-Time Time Hand Gesture Recognition Using Deep Learning Techniques. *Sensors* **2020**, *20*, 6451.
21. Günaydin, T.; Arslan, R. LOWER-LIMB FOLLOW-UP: A Surface Electromyography Based Serious Computer Game and Patient Follow-Up System for Lower Extremity Muscle Strengthening Exercises in Physiotherapy and Rehabilitation. In Proceedings of the 2019 IEEE 32nd International Symposium on Computer-Based Medical Systems (CBMS), Cordoba, Spain, 5–7 June 2019; pp. 507–512.
22. Iglesia, D.; Mendes, A.S.; Villarrubia, G.; Jiménez-Bravo, D.M.; Paz, J.F.D. Connected Elbow Exoskeleton System for Rehabilitation Training Based on Virtual Reality and Context-Aware. *Sensors* **2020**, *20*, 858. [CrossRef]
23. Aung, Y.M.; Al-Jumaily, A. *Augmented Reality Based Reaching Exercise for Shoulder Rehabilitation*; The Singapore Therapeutic, Assistive & Rehabilitative Technologies (START) Centre: Singapore, 2011.
24. Avila, A.; Chang, J.Y. EMG onset detection and upper limb movements identification algorithm. *Microsyst. Technol.* **2014**, *20*, 1635–1640. [CrossRef]

25. Balasubramanian, S.; Klein, J.; Burdet, E. Robot-assisted rehabilitation of hand function. *Curr. Opin. Neurol.* **2010**, *23*, 661–670. [CrossRef]

26. Lambercy, O.; Ranzani, R.; Gassert, R. Robot-assisted rehabilitation of hand function. In *Rehabilitation Robotics*; Elsevier: Amsterdam, The Netherlands, 2018; pp. 205–225.

27. Dovat, L.; Lambercy, O.; Gassert, R.; Maeder, T.; Milner, T.; Leong, T.C.; Burdet, E. HandCARE: A cable-actuated rehabilitation system to train hand function after stroke. *IEEE Trans. Neural Syst. Rehabil. Eng.* **2008**, *16*, 582–591. [CrossRef]

28. Hesse, S.; Schulte-Tigges, G.; Konrad, M.; Bardeleben, A.; Werner, C. Robot-assisted arm trainer for the passive and active practice of bilateral forearm and wrist movements in hemiparetic subjects. *Arch. Phys. Med. Rehabil.* **2003**, *84*, 915–920. [CrossRef]

29. Hesse, S.; Kuhlmann, H.; Wilk, J.; Tomelleri, C.; Kirker, S.G. A new electromechanical trainer for sensorimotor rehabilitation of paralysed fingers: A case series in chronic and acute stroke patients. *J. Neuroeng. Rehabil.* **2008**, *5*, 1–6. [CrossRef]

30. Lambercy, O.; Dovat, L.; Gassert, R.; Burdet, E.; Teo, C.L.; Milner, T. A haptic knob for rehabilitation of hand function. *IEEE Trans. Neural Syst. Rehabil. Eng.* **2007**, *15*, 356–366. [CrossRef] [PubMed]

31. Metzger, J.C.; Lambercy, O.; Gassert, R. Performance comparison of interaction control strategies on a hand rehabilitation robot. In Proceedings of the 2015 IEEE International Conference on Rehabilitation Robotics (ICORR), Singapore, 11–14 August 2015; pp. 846–851.

32. Metzger, J.C.; Lambercy, O.; Califfi, A.; Conti, F.M.; Gassert, R. Neurocognitive robot-assisted therapy of hand function. *IEEE Trans. Haptics* **2013**, *7*, 140–149. [CrossRef] [PubMed]

33. Kazemi, H.; Kearney, R.; Milner, T. Characterizing coordination of grasp and twist in hand function of healthy and post-stroke subjects. In Proceedings of the 2013 IEEE 13th International Conference on Rehabilitation Robotics (ICORR), Seattle, WA, USA, 24–26 June 2013; pp. 1–6.

34. Ball, S.J.; Brown, I.E.; Scott, S.H. A planar 3DOF robotic exoskeleton for rehabilitation and assessment. In Proceedings of the 2007 29th Annual International Conference of the IEEE Engineering in Medicine and Biology Society, Lyon, France, 23–26 August 2007; pp. 4024–4027.

35. Birch, B.; Haslam, E.; Heerah, I.; Dechev, N.; Park, E. Design of a continuous passive and active motion device for hand rehabilitation. In Proceedings of the 2008 30th Annual International Conference of the IEEE Engineering in Medicine and Biology Society, Bourke, AK, Canada, 20–24 August 2008; pp. 4306–4309.

36. Connelly, L.; Jia, Y.; Toro, M.L.; Stoykov, M.E.; Kenyon, R.V.; Kamper, D.G. A pneumatic glove and immersive virtual reality environment for hand rehabilitative training after stroke. *IEEE Trans. Neural Syst. Rehabil. Eng.* **2010**, *18*, 551–559. [CrossRef] [PubMed]

37. Cruz, E.; Kamper, D. Use of a novel robotic interface to study finger motor control. *Ann. Biomed. Eng.* **2010**, *38*, 259–268. [CrossRef]

38. Vanoglio, F.; Bernocchi, P.; Mulè, C.; Garofali, F.; Mora, C.; Taveggia, G.; Scalvini, S.; Luisa, A. Feasibility and efficacy of a robotic device for hand rehabilitation in hemiplegic stroke patients: A randomized pilot controlled study. *Clin. Rehabil.* **2017**, *31*, 351–360. [CrossRef] [PubMed]

39. Heo, P.; Gu, G.M.; Lee, S.j.; Rhee, K.; Kim, J. Current hand exoskeleton technologies for rehabilitation and assistive engineering. *Int. J. Precis. Eng. Manuf.* **2012**, *13*, 807–824. [CrossRef]

40. Rowe, J.B.; Chan, V.; Ingemanson, M.L.; Cramer, S.C.; Wolbrecht, E.T.; Reinkensmeyer, D.J. Robotic assistance for training finger movement using a hebbian model: A randomized controlled trial. *Neurorehabilit. Neural Repair* **2017**, *31*, 769–780. [CrossRef]

41. Mehrholz, J.; Pohl, M.; Platz, T.; Kugler, J.; Elsner, B. Electromechanical and robot-assisted arm training for improving activities of daily living, arm function, and arm muscle strength after stroke. *Cochrane Database Syst. Rev.* **2015**, *2015*, CD006876. [CrossRef]

42. Gomez-Donoso, F.; Orts-Escolano, S.; Garcia-Garcia, A.; Garcia-Rodriguez, J.; Castro-Vargas, J.A.; Ovidiu-Oprea, S.; Cazorla, M. A robotic platform for customized and interactive rehabilitation of persons with disabilities. *Pattern Recognit. Lett.* **2017**, *99*, 105–113. [CrossRef]

43. Costa, A.; Martinez-Martin, E.; Cazorla, M.; Julian, V. PHAROS—PHysical assistant RObot system. *Sensors* **2018**, *18*, 2633. [CrossRef]

44. Escalona, F.; Martinez-Martin, E.; Cruz, E.; Cazorla, M.; Gomez-Donoso, F. EVA: EVAluating at-home rehabilitation exercises using augmented reality and low-cost sensors. *Virtual Real.* **2019**, *24*, 1–15. [CrossRef

45. Zimmermann, C.; Brox, T. Learning to Estimate 3D Hand Pose from Single RGB Images. *CoRR* **2017**. [CrossRef]

46. Gomez-Donoso, F.; Orts-Escolano, S.; Cazorla, M. Accurate and efficient 3D hand pose regression for robot hand teleoperation using a monocular RGB camera. *Expert Syst. Appl.* **2019**, *136*, 327–337. [CrossRef]

47. Nasri, N.; Orts-Escolano, S.; Gomez-Donoso, F.; Cazorla, M. Inferring Static Hand Poses from a Low-Cost Non-Intrusive sEMG Sensor. *Sensors* **2019**, *19*, 371. [CrossRef] [PubMed]

48. Kim, M.; Cho, J.; Lee, S.; Jung, Y. IMU Sensor-Based Hand Gesture Recognition for Human-Machine Interfaces. *Sensors* **2019**, *19*, 3827. [CrossRef] [PubMed]

49. Neuron Motion Capture. Available online: https://neuronmocap.com/ (accessed on 8 March 2021).

50. Manus VR. Available online: https://www.manus-vr.com/ (accessed on 8 March 2021).

51. Leap Motion Controller. Available online: https://developer.leapmotion.com/ (accessed on 8 March 2021).

52. Weichert, F.; Bachmann, D.; Rudak, B.; Fisseler, D. Analysis of the Accuracy and Robustness of the Leap Motion Controller. *Sensors* **2013**, *13*, 6380–6393. [CrossRef]

53. Jin, H.; Chen, Q.; Chen, Z.; Hu, Y.; Zhang, J. Multi-LeapMotion sensor based demonstration for robotic refine tabletop object manipulation task. *CAAI Trans. Intell. Technol.* **2016**, *1*, 104–113. [CrossRef]
54. Gomez-Donoso, F.; Orts-Escolano, S.; Cazorla, M. Large-scale multiview 3D hand pose dataset. *Image Vis. Comput.* **2019**, *81*, 25–33. [CrossRef]
55. Li, G.; Li, Y.; Zhang, Z.; Geng, Y.; Zhou, R. Selection of sampling rate for EMG pattern recognition based prosthesis control. In Proceedings of the 2010 Annual International Conference of the IEEE Engineering in Medicine and Biology, Buenos Aires, Argentina, 31 August–4 September 2010; pp. 5058–5061.
56. Salvador, S.; Chan, P. Toward Accurate Dynamic Time Warping in Linear Time and Space. *Intell. Data Anal.* **2007**, *11*, 561–580. [CrossRef]
57. Bellman, R.; Kalaba, R. On adaptive control processes. *IRE Trans. Autom. Control* **1959**, *4*, 1–9. [CrossRef]
58. Loeb, G.E. Learning to Use Muscles. *J. Hum. Kinet.* **2021**, *76*, 9–33. [PubMed]
59. Parziale, A.; Senatore, R.; Marcelli, A. Exploring speed–accuracy tradeoff in reaching movements: A neurocomputational model. *Neural Comput. Appl.* **2020**, *32*, 13377–13403. [CrossRef]
60. Ning, L.; Manzhao, H.; Chuanxin, M.N.; He, C.; Y, W.; Ting, Z.; Peng, F.; Chih-hong, C. Next Generation Prosthetic Hand: FromBiomimetic to Bio-Realistic. *Research* **2020**, *2020*, 4675326.

Article

Intelligent Monitoring Platform to Evaluate the Overall State of People with Neurological Disorders

Jose Maria Vicente-Samper [1,*], Ernesto Avila-Navarro [2], Vicente Esteve [3] and Jose Maria Sabater-Navarro [1]

1   Department of Systems Engineering and Automation, Miguel Hernández University of Elche, 03202 Elche, Spain; j.sabater@umh.es
2   Department of Materials Science, Optics and Electronic Technology, Miguel Hernández University of Elche, 03202 Elche, Spain; eavila@umh.es
3   Department of Software and Computing Systems, University of Alicante, 03690 San Vicente del Raspeig, Spain; vesteve@ua.es
*   Correspondence: jose.vicentes@umh.es

Abstract: The percentage of people around the world who are living with some kind of disability or disorder has increased in recent years and continues to rise due to the aging of the population and the increase in chronic health disorders. People with disabilities find problems in performing some of the activities of daily life, such as working, attending school, or participating in social and recreational events. Neurological disorders such as epilepsy, learning disabilities, autism spectrum disorder, or Alzheimer's, are among the main diseases that affect a large number of this population. However, thanks to the assistive technologies (AT), these people can improve their performance in some of the obstacles presented by their disorders. This paper presents a new system that aims to help people with neurological disorders providing useful information about their pathologies. This novelty system consists of a platform where the physiological and environmental data acquisition, the feature engineering, and the machine learning algorithms are combined to generate customs predictive models that help the user. Finally, to demonstrate the use of the system and the working methodology employed in the platform, a simple example case is presented. This example case carries out an experimentation that presents a user without neurological problems that shows the versatility of the platform and validates that it is possible to get useful information that can feed an intelligent algorithm.

Keywords: sensors; electronic platform; machine learning; wearables

Citation: Vicente-Samper, J.M.; Avila-Navarro, E.; Esteve, V.; Sabater-Navarro, J.M. Intelligent Monitoring Platform to Evaluate the Overall State of People with Neurological Disorders. *Appl. Sci.* **2021**, *11*, 2789. https://doi.org/10.3390/app11062789

Academic Editor: Juan Antonio Corrales Ramon

Received: 24 February 2021
Accepted: 17 March 2021
Published: 20 March 2021

Publisher's Note: MDPI stays neutral with regard to jurisdictional claims in published maps and institutional affiliations.

1. Introduction

About one billion people live with some kind of disability. This corresponds to around 15% of the world's population [1]. The rate of disability is increasing, among other things due to the growth in the average age of the population and the increase in chronic health disorders. Lower income countries have a higher prevalence than higher income countries. Poor people have less resources to access treatment.

Half of people with disabilities cannot afford necessary medical care, compared to one third of non-disabled people who cannot afford it. People with disabilities have more than twice the probability of finding inadequate health care providers techniques, up to four times as likely to report improper treatment, and nearly three times as likely to be denied medical care [1].

Children with disabilities or disorders have a lower probability of attending school and receiving an adequate education. The probability of finding a job for a disabled person is also lower. Global employment data show that the employment rate for people with disabilities is 53% for men and 20% for women, while for people without disabilities it is 65% and 30% [1]. Therefore, people with disabilities are more vulnerable to poverty. They have worse living conditions due to the additional costs of their special needs (specialized medical care, assistive devices, or people for supporting them). As a result, people with disabilities are generally poorer than people without disabilities who have similar incomes.

Neurological disorders are diseases that affect the central and peripheral nervous system, i.e., the brain, the nerves that are found in the human body, and the spinal cord. Millions of people worldwide suffer from neurological disorders. For example, more than 6 million people die each year from strokes, more than 50 million people around the world have epilepsy, and 7.7 million cases per year are diagnosed with Alzheimer's disease, which is the most common cause of dementia [2]. The specific causes of neurological problems vary, but they may include genetic disorders, congenital anomalies or disorders, infections, lifestyle or environmental health problems, brain injuries, spinal cord injuries, or nerve injuries [3].

On the other hand, neurological disabilities include a wide range of disorders, such as epilepsy, learning disabilities, neuromuscular disorders, autism spectrum disorder, attention deficit disorder, brain tumors and cerebral palsy, among many others. Some neurological pathologies are cognitive, and they appear before birth. Other neurological disorders may be caused by tumors, degeneration, trauma, infection, or structural defects. Regardless of the cause, all neurological disabilities are the result of damage to the nervous system [3]. The need for technological solutions that help people who are affected by these pathologies becomes evident.

Assistive technologies (AT) are devices or systems that can be used to help a person with a disability or disorder to perform daily life activities. The AT can help to improve the functional independence and thus, facilitate the daily living tasks through the use of aids that help a person to travel, to communicate with others, to learn, to work, and to participate in social and recreational activities [4]. AT devices can range from a simple and with low technology designs such as a crutch, to complex systems that speak for the user, automatic opening doors systems or to brain-wave recognition units for interface management [5].

Some of these assistive devices try to manage the associated problems of neurological disorders and facilitate the daily life of both the people who suffer from them and the family or support members who help them throughout their lives. An example is the Embrace device from Empatica [6]. It is a wrist-worn wearable device that monitors the user to detect possible convulsive seizures and alert caregivers. Another example of a monitoring device is the PdMonitor® from PD Neurotechnology [7]. It is a set of wearable monitoring devices, in this case for people with Parkinson's disorder. The device tracks, records, and processes a variety of symptoms, often present in this disease. A third example is the Monarch eTNS® System from NeuroSigma [8]. It is the first device approved by the U.S. Food and Drug Administration (FDA) for the treatment of the Attention Deficit disorder in children (ADHD). The device sends a low-level electrical pulse through a wire to a small patch adhered to the patient's forehead. The therapeutic pulses stimulate the branches of the trigeminal nerve, which activates the neural pathway to other parts of the brain thought to be involved in ADHD. Many research works for the development of new assistive devices can also be found. For example, the work of Cesareo et al. (2020) presents a system for monitoring the breathing rate in people with muscular dystrophy [9]. The system consists of a set of inertial measurement units integrated in wearable devices to control the long-term breathing pattern. Another work in progress is Floodlight Open, developed by the Hoffmann-La Roche company [10]. This is a study that aims to monitor the multiple sclerosis (MS) symptoms using a smartphone, through simple tasks specifically designed to assess the effects of MS.

There are other assistive devices that could be defined as intelligent platforms. These modular systems, in addition to monitoring the user and providing information about the user's condition, incorporate some type of intelligence in the form of decision algorithms or new machine learning methods that improve more traditional systems. For example, an intelligent tool for assisting people with Alzheimer's disease is presented in [11]. The system helps to monitor the user's health, control medication or locate the user when they become disoriented, among other things. The system is composed by multiple devices that monitor the user, records their position, and control medication and objects that may be important. Through a mobile application, the user and caregivers have access to information and receive alerts. Another idea is presented in [12]. It proposed using wearable monitoring devices together with computational intelligence to diagnose and monitor people with Parkinson's disease. The assessment of Parkinson's Disease motor disabilities is based on neurological examination during patient's visits to the clinic and home diaries that the patient or the caregiver keeps. However, the short-time examination may not reveal important information. To overcome these limitations and difficulties, the ambulatory monitoring systems can improve this evaluation. Applying machine learning algorithms to these platforms allows to obtain intelligent systems for assistance. For example, Casalino et al. (2018) [13] present a system for real-time monitoring of cardiovascular problems using video images and fuzzy inference rules. The proposed system is composed of a transparent mirror with a camera that detects the user's face. The frames are processed using photoplethysmography in order to estimate different physiological parameters of the user. The estimated parameters are used to predict a risk of cardiovascular disease through fuzzy inference rules. Another system that employs the use of monitoring devices and machine learning algorithms is [14]. The authors present a gait-assistive system using a neural network. The system is composed of devices that monitor the user's movement during gait and stimulate the muscle nerves using electrical stimulation through electrodes. After a data collection phase, a model based on recurrent neural networks is trained. The model will be in charge of predicting the user's movement during gait and controlling the stimulation signal.

The purpose of this work is to present a full platform for the development of custom predictive models that help people with neurological disorders. Figure 1 sketches the concept of this work. The main challenges that this work faces are the signal acquisitions from the user and the environment, the signal processing, the dataset generation with feature engineering, and to train and optimize a predictive model. The aim is to use the generated model to help the user to manage their pathologies; therefore, the model will become an AT in his daily life. Unlike other works where the system is focused on a specific pathology or the algorithms used are optimized for a specific application, the proposed system is intended to be used for multiple pathologies and applications. The platform is presented in a generic way, where each of the stages can be adapted in order to obtain a different predictive model. This model will be personalized for the user and the desired application. Thus, to illustrate the versatility of the platform and to show the working methodology, a system validation experimentation is performed.

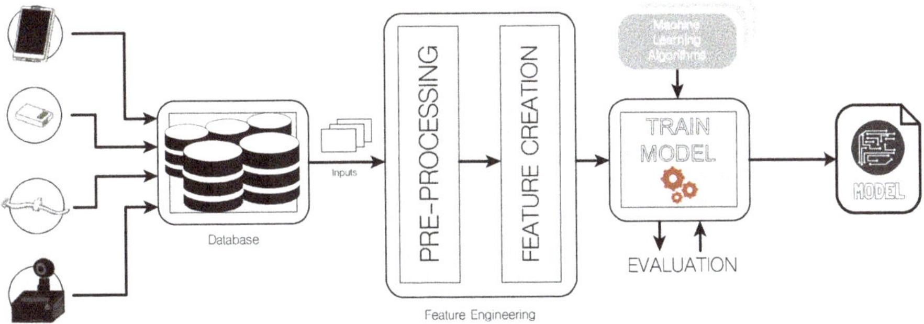

Figure 1. General concept of the proposed platform.

This paper is organized as follows. In the materials and methods section, the description of the different parts that make up the platform is presented. First, the acquisition system composed by four modular electronic devices. Then, Sections 2.2 and 2.3 describe the characteristics of the database and the feature engineering before training the model, respectively. Section 2.4 shows the steps to follow towards generating predictive models using machine learning (ML) algorithms. In Section 2.5, a use case of the platform is presented; it is intended to generate a prediction model for people concentration in the workstation. Section 3 shows the results of the use case with data extracted from the generated model and its training. Finally, in Sections 4 and 5 the obtained results are discussed and the conclusions of the paper are outlined.

2. Materials and Methods

This section presents a platform for the generation of predictive models for people with neurological disorders (Figure 1). These models, customized for each user, intend to provide information about the pathology of the user and to help them manage it in a more controlled way. The generated information can also be used as input or as feedback for other assistive technologies. First, the different parts that make up the platform are described. Each of the stages can be adapted to the special conditions of the user and the final objective of the application to be developed. To conclude the section, an example of the use of the platform is described to show the workflow of the system.

2.1. Data Acquisition System

The first stage of the platform is the data acquisition system. This system is responsible for obtaining information from the user and from the surrounding environment. It has been developed with a modular architecture that allows adjusting the use of the devices according to the special characteristics of each user and the application that is carried out. For example, it allows the sensors to be restructured in the event that one of the devices is no longer required. An example of this would be to integrate the sound sensor into the smartphone in situations where the video device is not used, or vice versa. The system consists of four electronic devices: a smartphone, a wrist wearable device, an environmental monitoring device, and a video sensor device.

2.1.1. Personal and Environmental Devices

The environmental and personal monitoring devices are responsible for measuring the environmental conditions and the user's physiological variables, respectively. The development of these devices, which are integrated within the presented platform in this work, along with a study of the different parameters measured by this data acquisition system is described in [15].

The environmental monitoring device provides information about the luminosity, the environmental temperature, the relative humidity, and the atmospheric pressure of the environment where the user is. It is a small electronic device, which could be a key ring.

The personal monitoring device is the one described in [15], which integrates the measurement of the heart rate and the body temperature of the user. To complement the device, a new sensor has been integrated. It is an Inertial Measurement Unit (IMU), which provides information about the motor activity performed by the user. The details of this sensor and its integration into the system are described in Section 2.1.2. This personal monitoring equipment consists of a small wearable wrist device. Its design has been slightly updated to integrate the new sensor, and its dimensions have been reduced with respect to the previous version, despite integrating a larger capacity battery that provides an autonomy of more than 20 h of use. In addition, it has been manufactured with soft and comfortable materials that provide ergonomics and facilitate its placement for users with special difficulties. Figure 2 shows a picture of the new design, where the device, completely covered with EVA foam, can be seen. This also allows the device to weigh only 5.90 g, including the foam cover.

Figure 2. New design of the personal monitoring device.

On the other hand, the management of these monitoring devices is done with a smartphone via an Android application. The application, also described in [15], also allows displaying information to the user or to the caregiver. The interface and the displayed information can be modified depending on the user and the corresponding application. Furthermore, as described in [15], the smartphone also acts as a sound sensor, capturing the entire human audible spectrum (between 20 Hz and 22 kHz).

2.1.2. Motor Activity Sensor

The motor activity (MA) of the user is a relevant parameter that provides useful information to the platform. For example, it allows to evaluate the health and wellness in users with neurodegenerative disorders that influence in the motor functionality [16]. This parameter allows to quantify the arm movements and provides data about the user's displacement. In addition, together with other parameters, it is possible to estimate the state of mind or measure stress levels of the user [17].

The sensor used for the measurement consists of an IMU placed in the personal monitoring device, which integrates 3-axis gyroscope, 3-axis accelerometer, and 3-axis magnetometer. As an example, Figure 3 shows a fragment of the obtained signals by the MA sensor in the X-axis during a test session.

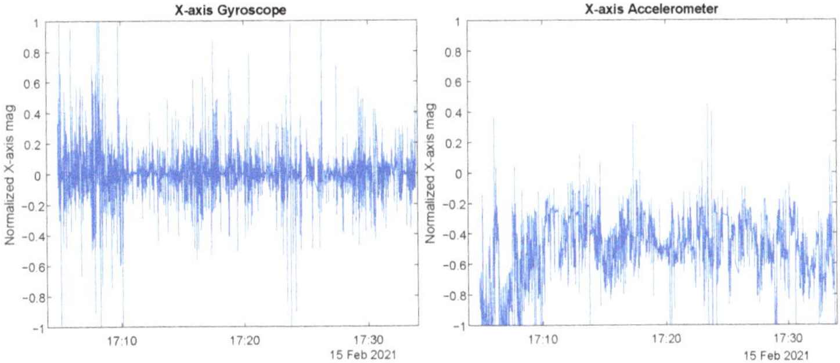

Figure 3. Example fragment of X-axis Gyroscope and Accelerometer signals recorded from the MA sensor during a test session.

The obtained data every sample by the sensor are organized in an array and stored in the corresponding collection within the database. The structure of a sample document for storing samples is shown below. The *date* and *dateString* parameters are the timestamp of the measurement in milliseconds and in character string, respectively. These parameters are included in every document in all data collections. The generated array with the measurement data is stored in the *motor* parameter. The *ObjectId* identifier is automatically set by the database when the document is uploaded.

```
{
  "_id" : ObjectId
  "date" : NumberLong
  "dateString" : "EEE MMM d HH:mm:ss z yyyy"
  "motor" : [gyrX, gyrY, gyrZ, accX, accY, accZ, magX, magY, magZ]
}
```

2.1.3. Video Device

The last device that makes up the data acquisition stage is the video device. The purpose of using this system is to obtain visual relevant information to the applications from the user's environment.

Some neurological disorders can cause difficulties in social interaction and therefore affect the mental wellbeing of the user [18]. Thus, one of the parameters measured by the video device is the number of people in the user's environment at any given time. In this way, the platform has information which together with other parameters allows to relate, for example, how social interaction can affect the user or whether an excessive presence of people around the user influences in the user's behavior [19].

On the other hand, regardless of the number of people around the user, some people with neurological disorders may feel overwhelmed if there is excessive activity around them, such as constant moving around or movements close to the user. In addition, these actions added to other stimuli, e.g., ambient noise, may be intensified [20]. Therefore, another parameter provided by the video device to the platform is the Optical Flow. The Optical Flow is the apparent motion pattern of objects in the image between two consecutive frames, caused by the displacement of the object or the camera. It is a 2-dimensional vector field where each vector represents a displacement vector indicating the movement of a point from its position in the first frame to its position in the second frame [21]. Consequently, if we keep the camera in a fixed position, the displacement that occurs in the image between frames will then be due only to the movement of the objects. In other words, we will have an estimate of how much movement occurs in the user's environment. The video device consists of an Nvidia Jetson Nanocompute card [22] and an Insta360 Air 360 degree camera [23]. The Jetson Nano has 128 $CUDA^{®}$ (Compute Unified Device Architecture) cores that encourage the execution of applications where computer vision and machine learning algorithms are used. The 360 degree camera provides the most complete possible picture of the user's environment. The system is mounted in a $118 \times 96 \times 60$ mm aluminum casing for portability between rooms. In addition, an improved cooling kit has been added to keep the temperature stable during use. An image of the video device is shown in Figure 4a.

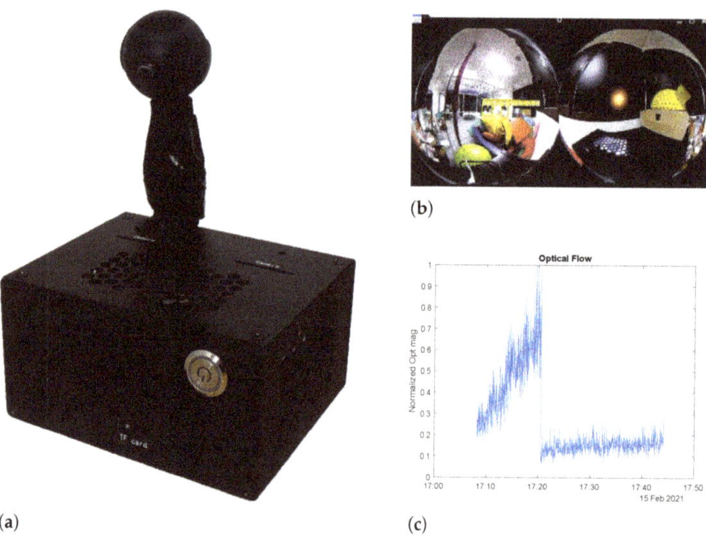

(a)   (b)   (c)

Figure 4. (a) Picture of the video device. (b) Screenshot of a people detection frame during a test session. (c) Example fragment of the optical flow signal recorded during a test session.

The first parameter introduced is the quantification of people in the user's environment. To do this, it is necessary to detect the people who are in the room at any given moment. This detection task is performed using a convolutional neural network (CNN) model. The algorithm used is YOLOv5 [24,25]. A pretrained model optimized for object detection is used. It has a high speed of execution doing inference and its size is contained. Figure 4b shows a screenshot of a frame where the detection of people during a test session is observed. It can be seen how the 360 degree camera captures the entire environment and the algorithm detects where the people are. The structure of a sample document for storing People Detection data is shown below. The *people* parameter stores the measured number of people.

```
{
  "_id" : ObjectId
  "date" : NumberLong
  "dateString" : "EEE MMM d HH:mm:ss z yyyy"
  "people" : NumberShort
}
```

On the other hand, there are different ways to measure the Optical Flow. In this case, a dense measurement of Optical Flow has been chosen, in which the displacement of all the points of the frame is calculated. To measure the Optical Flow of all the points in the image and to be able to quantify the movement that occurs in the user's environment, the method of Gunnar Farnebäck is used [26]. This algorithm is based on polynomial expansion and performs an estimation of the motion of two frames. Finally, the displacement modulus of both axes (X and Y) of the frame is stored. Figure 4c shows a fragment of the Optical Flow signal recorded during a test session. It can be observed how after a period of time the signal level is lower. This fact is due to a change of position by the people in the session to a place farther away from the camera, which results in a smaller change between frames and therefore a lower signal magnitude. The structure of a sample document for storing Optical Flow data is shown below. The *optical* parameter stores the estimated optical flow.

```
{
  "_id" : ObjectId
  "date" : NumberLong
  "dateString" : "EEE MMM d HH:mm:ss z yyyy"
  "optical" : NumberLong
}
```

2.2. Database

The next stage of the platform is in charge of storing the information obtained by the data acquisition system. The information must be organized in an efficient and secure way. Therefore, the database must have some characteristics that allow the platform to work properly and with minimum execution times. The main properties that must be met are listed below.

- It must allow the integration of multiple types of data, as well as change the architecture in which the information is stored if necessary.
- It must have continuous availability and scalability.
- It must support time-series. This requires less storage space and allows for faster query speed.
- It must perform continuous data updates that allow real-time analysis of the information.
- It must have strong access controls, data audits, and protect the data with encrypted controls. The information stored in the database is very valuable to the user and to the platform itself. Thus, the security standards must be met to ensure data anonymity and data access restriction.

The database used in the platform is a non-relational MongoDB database [27]. The information obtained by the data acquisition system is stored in documents within data collections, with a collection for each measured parameter. This allows new parameters to be integrated into the platform without the need to modify the storage architecture. An example of the collection storage structure is shown in Figure 5.

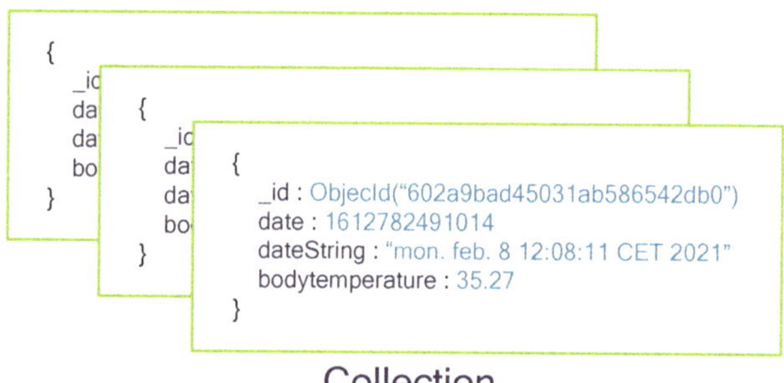

Figure 5. Example of database collection in the platform.

2.3. Feature Engineering

Feature engineering is the process of extracting attributes that improve the machine learning model from the raw information using data mining techniques [28]. The features must have an appropriate configuration. Therefore, these features are created from the previously acquired data, and their inputs are transformed and prepared for the machine learning (ML) model. Synthetic features that do not originally exist in the dataset are also created from the available information and they will allow the model to perform better.

The features have to comply with a series of properties in order to be suitable for use in the model. The attribute must be related to the objective of the application, i.e., it must be relevant in the output that our model is looking for. In addition, it must be certain that it will be possible to know the value of that feature at the time of the prediction; otherwise, the model will not work correctly. The value of the features must be numerical and must represent a magnitude. This is because, for example, a neural network is no more than a machine that performs arithmetic, trigonometric, and algebraic operations on the input variables [29]. Therefore, those features that are not numerical but are required to be used will have to be converted or encoded before introducing them into the model. Finally, it is also necessary to have enough examples of the features values to train the model correctly. The above rules are used to extract the appropriate features for the application and to generate a dataset that will be used to train and to validate the desired model. The transformations to be performed on the available signals in order to extract the attributes will depend on the final objective of the application. Therefore, once the problem to be faced has been defined, the signals must be analyzed and those that will be useful for the model must be selected. Subsequently, the necessary modifications will made.

The final aim of this stage is to generate a suitable dataset to train the model. As mentioned above, this stage depends entirely on the information obtained from the model. Therefore, it is directly related to the result of the model training and could be placed in a parallel position to the next stage. In most of the times, it will be necessary to modify the dataset with new information or to change the existing features.

2.4. Training and Generation of a Custom Model

Using the generated feature dataset from the information collected by the data acquisition system, a customized model is trained for the user and the desired application. For this purpose, appropriate machine learning algorithms are used according to the specific application, i.e., it is particularized according to the information that will be predicted in the future. The training and evaluation process of the model is performed cyclically until the optimal and sufficiently accurate behavior of the model is found. The dataset will be modified using the information obtained during training to improve the model accuracy. Finally, a validation of the model is performed with a part of the dataset reserved for testing. This allows checking the predictive behavior of the trained model, analyzing possible lines of improvement in the process, and verifying the accuracy for new inputs to the platform.

Once the training, validation, and testing processes of the model are completed, the trained model can be used to predict information from new inputs to the platform taken with the acquisition system. The model can be hosted on a client that runs continuously and can provide real-time feedback to the user.

The first step to be performed before starting the training is the split of the dataset into three parts. A diagram of this split is shown in Figure 6. The first fragment will be the training data (training set). The training set is the dataset used by the model to learn how to process the information. From these data, the model adjusts the parameters of the classifier or from the algorithm used in the model. Using the training set, different machine learning algorithms can be evaluated to generate the model, and the results obtained can be compared to obtain the most appropriate one for the application. The training set contains most of the data from the main dataset. The validation set is used to estimate how well the model has been trained. This can be done between periodic training cycles and when the training process has been completed. It is also used to estimate model properties, such as the error in a classifier or the precision and recall in binary models. Cross-validation is the most commonly used method for this task [30]. Finally, there is the test set. It is used only to evaluate the performance of the model after the training process has been completed. It can be considered as a mock production use of the model.

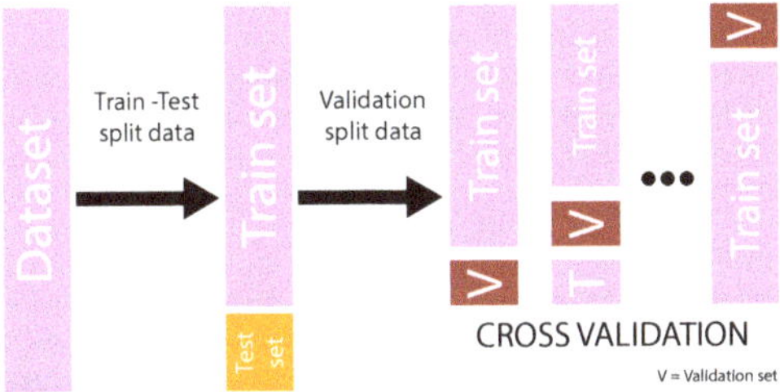

Figure 6. Workflow of dataset split and cross validation process.

2.5. Creation of a People Concentration Model in the Workstation

The system presented in this paper can be used for multiple pathologies and applications. The acquisition system and the database allow to easily adapt or integrate new sensors. The signal processing, the feature analysis, and the machine learning algorithms used to generate the model will also be defined according to the objective. To show the workflow of the platform, an experimentation to validate the system is performed. The test consist of generating a personalized model of the concentration of a person in two different workstations within the same research laboratory. One of the places is located at the entrance of the laboratory, next to the door. Any person who must access or leave the laboratory has to pass by this position, which we will call Position A. The second position (Position B) is located in an isolated room within the same laboratory.

The final goal of the model is to predict at which times the person is focused on the tasks being performed and at which times the concentration is reduced, for example, due to an interruption or stimuli around the workstation. This will provide information on which of the positions is the most suitable to work at. Although this application is not directly related to assisting people with neurological disorders, it is an example of the versatility that the system can offer as well as to show the working methodology of the platform. In addition, a similar model could also be proposed for a person with a neurological disorder so that it would be possible to analyze how the user adapts to a specific work environment. The two possible situations are labeled as *"Focused"* and *"Distracted"*, and the output of the predictive model will be one of these two options.

First, it is necessary to define what will be considered as *"Focused"* and *"Distracted"*. For this example, a first supervised part has been proposed where data are collected from the person and the environment in both workstations performing a task that allows to quantify the person's concentration on the task. This task consists of reading an entertainment book. The reading of technical or scientific documents, which a priori could be more related to a task to be developed in the person's work, has been discarded, because the reading pace may not be constant due to the complexity of certain parts throughout the document. During this supervised period, the number of lines per minute read by the user is counted. In this way, a range is established in which the person is considered focused on the task. Below that range, the person is considered distracted. The supervised task is necessary to establish the user's concentration level in a quantitative way. The model needs this information to learn in which situations the user is focused on the task.

Once the supervised stage is completed, the necessary information is available to generate the new model, always based on the previously established definition of concentration. The second part of the experimentation corresponds to the unsupervised collection of new data in both workstations. The user performs the usual tasks of his job. The new information collected will be used to estimate the concentration levels of the person in both workstations using the generated model with the supervised data.

For the supervised stage, a data acquisition period of 180 min has been established for each of the established positions (A and B), i.e., a total of 6 h. To make the supervised experimentation more user-friendly, the time has been divided into three one-hour periods, taken on consecutive days. Subsequently, for the unsupervised stage, a total of 8 h of data have been collected. For each of the positions, 4 h have been taken. In turn, these 4 h have been captured in 2-h periods on consecutive days.

The next step is to analyze the available signals and decide which data can provide relevant information to the model. As the experimentation is performed in a controlled environment, the weather conditions are kept constant over time at both workstations. Therefore, ambient temperature, relative humidity, and atmospheric pressure do not provide information to the model that helps to predict the concentration level and are discarded. Similarly, the use of the ambient luminosity can be discarded as the values also remain constant over time and of the same magnitude, both at position A and B. Finally, the use of the magnetometer signal from the MA sensor is also discarded because it is not necessary to know the user's position. Thus, the heart rate and the body temperature signals of the user, the gyroscope and the accelerometer signals of the MA sensor, the ambient sound, and the signals of the video device (optical flow and people detection) would remain.

When the available signals to be used have been decided, a preprocessing is performed before generating the dataset. The different transformations applied to each of them are described below.

- **Sound:** The audio spectrum is available divided into frequency bands. For the desired model, the information isolated by frequencies is not needed. Thus, a feature that contains the accumulated energy value in all frequency bands is generated. We use Equation (1) for this purpose:

$$Sound_{total} = 10log\left(\sum_{i=1}^{n} 10^{\frac{Sound_i}{10}}\right) \qquad (1)$$

- **Heart Rate:** The heart rate value is available in 6-s windows. As the photoplethysmography signal used to obtain the heart rate value is very sensitive to motion artifacts, the 75th percentile of the last 30 s is used as feature for the dataset.
- **Body Temperature:** The body temperature value does not require any special transformation. The obtained signal itself is used as a feature.

- **Gyroscope and Accelerometer:** The gyroscope and the accelerometer signals are available separately in the three axes (X, Y, and Z). For the desired model the divided information is not needed, so a feature is generated with the module of the three axes for the gyroscope signal and another one for the accelerometer signal.
- **Optical Flow and People Detection:** Both the Optical Flow signal and the quantification of people do not require any special transformation for the dataset. The signals obtained are used as features.

Finally, the signals of all the generated features are normalized, excluding the quantification of people. A fragment of these signals can be seen in the Appendix A in Figure 1. Once the features have been established and the different transformations have been carried out, the dataset is generated. For this purpose, a time vector is created where the time values of all the features are chronologically ordered together inside with an identifier that indicates the feature to which they belong. In addition, an index is assigned to each of the features indicating the last entry added to the dataset. Then, the time vector is browsed, adding to the dataset the values of the different features of the position indicated by each of its indexes. This merging generates an input in the dataset. Once the entire time vector has been browsed, the dataset is complete with all the information generated. To describe this process graphically, a schematic is shown in Figure 7.

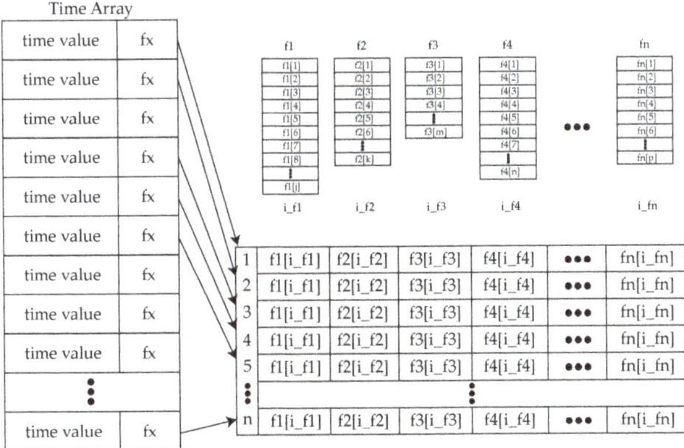

Figure 7. Diagram of the dataset generation process.

Using the described method, both datasets are generated for our model with the data collected in the experiment (supervised set and unsupervised set). Table 1 shows the distribution of both. Therefore, joining the records in both positions (A and B), we have a dataset for training and validation of the model with 17,135 entries, with 34.08% of entries labeled as *"Distracted"*.

Table 1. Datasets distribution.

	Supervised Data			
	Position A		Position B	
Class	**Records**	**Percentage**	**Records**	**Percentage**
Focused	5354	62.18%	5942	69.71%
Distracted	3257	37.82%	2582	30.29%
	Unsupervised Data			
	Position A		Position B	
Records	11350		11234	

For training and validation of the model, the dataset has been split (80–20%). Between different checked algorithms, a sequential neural network model has been selected due to its performance with the available dataset. The neural network architecture has also been adjusted from the simplest to obtain an adequate performance without increasing the inference time excessively. The selected neural network consists of the input layer, two hidden layers with 120 and 80 nodes with *"relu"* type activation function and the output layer with 2 nodes and *"softmax"* as activation function. A schematic of the model structure is shown in Figure 8. The parameters used for training the model are as follows.

$$
\{
$$
$$
Optimizer = Adam,
$$
$$
Loss\ function = Sparse\ Categorical\ Crossentropy,
$$
$$
Batch\ size = 10,
$$
$$
Numer\ of\ epochs = 500
$$
$$
\}
$$

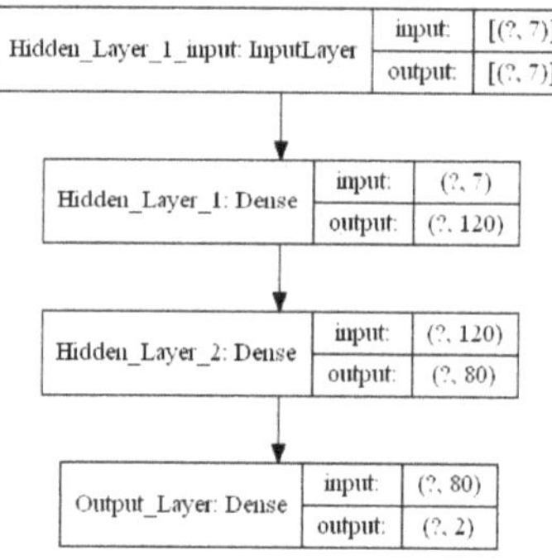

Figure 8. Architecture of the model.

3. Results

It should not be forgotten that the final aim of the experimentation is to generate a model of the concentration of a person in two different workstations within the same laboratory. This section attempts to present the results obtained from both the training of the model and the experimentation itself.

The obtained results of the model training are summarized in the Table 2. The Receiver Operating Characteristic (ROC) curve and its respective Area Under Curve (AUC) are included in Figure 9. The performance of the model is very fine. The model is able to recognize which dataset inputs have been defined as *"Focused"* and *"Distracted"* with high accuracy.

Table 2. Report of the model performance.

Class	Precision	Recall	F1-Score
Focused	0.97	0.95	0.96
Distracted	0.93	0.96	0.94
Average/Total	0.95	0.95	0.95
Accuracy		95.26%	

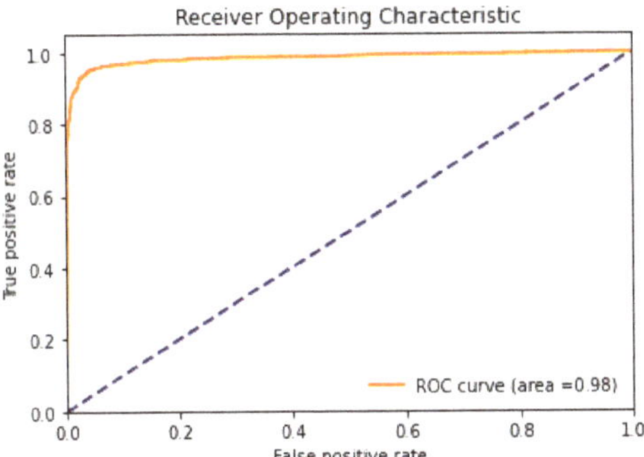

Figure 9. Receiver Operating Characteristic (ROC) curve of the model training result.

Once the model has been generated and evaluated its performance, the unsupervised dataset is used in order to identify each dataset input as a *"Focused"* or *"Distracted"* situation using the model. The results obtained are shown in Figure 10. A comparative graph between both positions (A and B) can be seen.

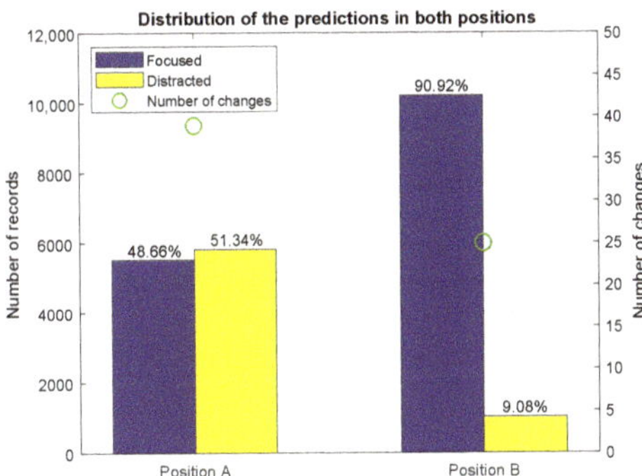

Figure 10. Obtained distribution of the predictions of the unsupervised data.

4. Discussion

A model with successful results has been trained only using seven generated features from all the integrated signals in the platform. As can be seen in Table 2, the error in the model evaluation is very low. The AUC is very close to 1 (Figure 9), which means that the obtained results in the evaluation are almost perfect.

Analyzing the generated predictions by the model from the unsupervised set (Figure 10), the total time of concentration in Position B is higher. There is a notable difference in the time that the user remains focused on his tasks in Position A (51.34%), with respect to Position B, where more than the 90% of the time has been identified as *"Focused"*. In addition, the concentration time has undergone a 50% more changes in the level of concentration in Position A (39) than in Position B (25). This implies that concentration periods are shorter between distractions. Therefore, it seems that as common sense would indicate, Position B is a better workstation.

Regarding the performance of the initial stages of the platform, the acquisition data system has allowed obtaining signals with good quality in all devices and the storage protocol has worked correctly. The analysis of the signals and the processing carried out to extract the features have been customized for the application that has been presented as an example of use. However, the platform presented in this work has demonstrated high versatility when it comes to generating new information that can help people with neurological disorders.

5. Conclusions

This work aims to present an intelligent platform that could provide useful information about people with neurological pathologies as an assistive technology. The novelty of this work is the acquisition of physiological and environmental signals for the generation of predictive models using machine learning algorithms. Throughout the paper, the different stages that make up the system have been described. Finally, an example of the use of the platform is presented, which allows a more detailed description of each of the steps taken until the desired model is obtained. The proposed application in the experimentation does not have a direct relationship with the assistance to people with neurological disorders. However, it has made it possible to describe the work at each stage of the platform. In addition, the presented example could be transformed into a real case with a user with a neurological disorder. For example, the platform could be used to measure the user's level of adaptation to a particular job.

This paper attempts to show the versatility offered by the presented system. The modular design allows to integrate or adapt different sensors. The analysis and processing of the signals will be defined according to the objective for which the platform is to be used. Likewise, the machine learning algorithms used will depend on the model to be obtained. However, to generate a real predictive model that helps to manage problems arising from a neurological disorder, it is necessary to collect a much larger amount of information than the used for the example case. In addition, the feature engineering that must be developed to obtain the appropriate features for the model is also very complex. Therefore, the system has been presented in a generic way together with a simple example to show the workflow.

Currently, the system is in use for the development of a personalized models of people with ASD, one for each user. Every model should predict behavioral changes in the user due to environmental stimuli, caused by a sensory processing problem derived from their pathology. Data collection is being developed in a clinical setting. Another possible use could be to obtain information about the stress level of a person with reduced mobility. The information provided by the platform could be used to modify the user's posture, alert the caregiver, etc. Different applications could also be to have an intelligent feedback system during a session of rehabilitation activities, or to be used in Sleep Disorders Units.

Author Contributions: Conceptualization, J.M.V.-S. and J.M.S.-N.; methodology, J.M.V.-S. and J.M.S.-N.; software, J.M.V.-S.; validation, J.M.S.-N., E.A.-N., and J.M.V.-S.; formal analysis, J.M.V.-S., V.E., and E.A.-N.; investigation, J.M.V.-S.; resources, J.M.V.-S. and E.A.-N.; data curation, J.M.V.-S.; writing—original draft preparation, J.M.V.-S.; writing—review and editing, J.M.S.-N., E.A.-N., V.E., and J.M.V.-S.; visualization, J.M.V.-S.; supervision, J.M.S.-N., E.A.-N., and J.M.V.-S.; project administration, J.M.S.-N. and J.M.V.-S.; funding acquisition, J.M.S.-N. and E.A.-N. All authors have read and agreed to the published version of the manuscript.

Funding: This work was partially funded by Spanish Research State Agency and European Regional Development Fund through "Race" Project (PID2019-111023RB-C32). The work of J.M.V.-S. is supported by the Conselleria d'Educació, Investigació, Cultura i Esport (GVA) through FD-GENT/2018/015 project.

Institutional Review Board Statement: Not applicable.

Conflicts of Interest: The authors declare no conflict of interest.

Appendix A

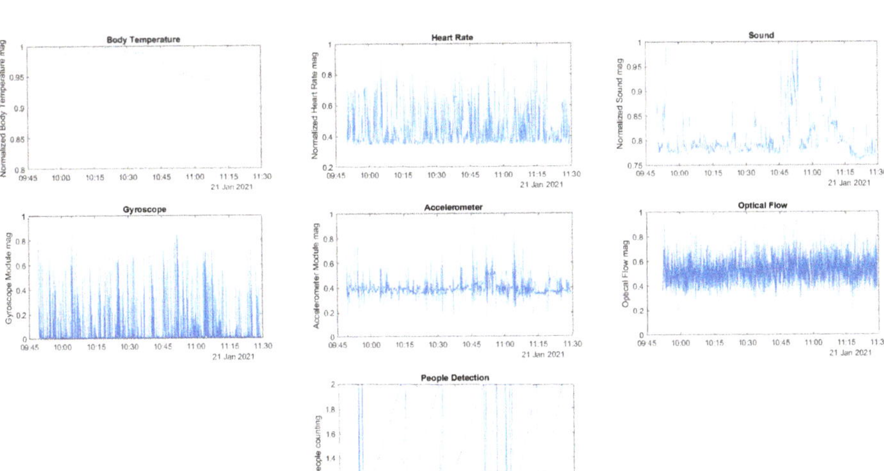

Figure 1. Fragment of the selected signals as features during the experimentation.

References

1. World Health Organization. Ten Facts on Disability. 2017. Available online: https://www.who.int/features/factfiles/disability/en/ (accessed on 5 February 2021).
2. World Health Organization. Mental Health: Neurological Disorders. 2016. Available online: https://www.who.int/features/qa/55/en/ (accessed on 5 February 2021).
3. Montana Department of Public Health & Human Services. Neurological Disorders. Available online: https://dphhs.mt.gov/schoolhealth/chronichealth/neurologicaldisorders (accessed on 5 February 2021).
4. Centers for Disease Control and Prevention. People with Disabilities. 2020. Available online: https://www.cdc.gov/ncbddd/disabilityandhealth/people.html (accessed on 5 February 2021).
5. Cook, A.M.; Polgar, J.M. Chapter 2—Technologies that Assist People Who Have Disabilities In *Assistive Technologies*, 4th ed.; Cook, A.M., Polgar, J.M., Eds.; Elsevier: Maryland Heights, MO, USA, 2015; pp. 16–39. [CrossRef]
6. Empatica. Embrace 2—Smarter Epilepsy Management. 2020. Available online: https://www.empatica.com/en-eu/embrace2/ (accessed on 6 February 2021).
7. PD Neurotechnology. PD Monitor Solution. 2021. Available online: https://www.pdneurotechnology.com/pd-monitor-solution/product/ (accessed on 6 February 2021).
8. NeuroSigma. Monarch eTNS System® for Pediatric ADHD. 2020. Available online: https://www.monarch-etns.com/ (accessed on 6 February 2021).
9. Cesareo, A.; Nido, S.A.; Biffi, E.; Gandossini, S.; D'angelo, M.G.; Aliverti, A. A wearable device for breathing frequency monitoring: A pilot study on patients with muscular dystrophy. *Sensors* **2020**, *20*, 5346. [CrossRef] [PubMed]
10. Hoffmann-La Roche. Floodlight Open. Understanding MS Together. 2021. Available online: https://floodlightopen.com/ (accessed on 6 February 2021).
11. Omar, K.S.; Rabbi, F.; Anjum, A.; Oannahary, T.; Karim Rizvi, R.; Shahrin, D.; Anannya, T.T.; Nasreen Tumpa, S.; Karim, M.; Nazrul Islam, M. An Intelligent Assistive Tool for Alzheimer's Patient. In Proceedings of the 1st International Conference on Advances in Science, Engineering and Robotics Technology (ICASERT), East West University, Dhaka, Bangladesh, 3–5 May 2019. [CrossRef]
12. Romero, L.E.; Chatterjee, P.; Armentano, R.L. An IoT approach for integration of computational intelligence and wearable sensors for Parkinson's disease diagnosis and monitoring. *Health Technol.* **2016**, *6*, 167–172. [CrossRef]
13. Casalino, G.; Castellano, G.; Pasquadibisceglie, V.; Zaza, G. Contact-less real-time monitoring of cardiovascular risk using video imaging and fuzzy inference rules. *Information* **2018**, *10*, 9. [CrossRef]
14. Tang, S.Y.; Hoang, N.S.; Chui, C.K.; Lim, J.H.; Chua, M.C.H. Development of Wearable Gait Assistive Device Using Recurrent Neural Network. In Proceedings of the 2019 IEEE/SICE International Symposium on System Integration (SII), Paris, France, 14–16 January 2019; pp. 626–631. [CrossRef]
15. Vicente-Samper, J.M.; Ávila-Navarro, E.; Sabater-Navarro, J.M. Data Acquisition Devices Towards a System for Monitoring Sensory Processing Disorders. *IEEE Access* **2020**, *8*, 183596–183605. [CrossRef]

16. Prathivadi, Y.; Wu, J.; Bennett, T.R.; Jafari, R. Robust activity recognition using wearable IMU sensors. In Proceedings of the IEEE Sensors, Valencia, Spain, 2–5 November 2014; pp. 486–489. [CrossRef]
17. Can, Y.S.; Arnrich, B.; Ersoy, C. Stress detection in daily life scenarios using smart phones and wearable sensors: A survey. *J. Biomed. Inform.* **2019**, *92*, 103139. [CrossRef] [PubMed]
18. Henry, J.D.; Von Hippel, W.; Molenberghs, P.; Lee, T.; Sachdev, P.S. Clinical assessment of social cognitive function in neurological disorders. *Nat. Rev. Neurol.* **2016**, *12*, 28–39. [CrossRef] [PubMed]
19. Altogether Autism. A Shift in Perspective: Empathy and Autism. 2021. Available online: https://www.altogetherautism.org.nz/a-shift-in-perspective-empathy-and-autism/ (accessed on 17 February 2021).
20. Clark, C.; Crumpler, C.; Notley, H. Evidence for environmental noise effects on health for the United Kingdom policy context: A systematic review of the effects of environmental noise on mental health, wellbeing, quality of life, cancer, dementia, birth, reproductive outcomes, and cognition. *Int. J. Environ. Res. Public Health* **2020**, *17*, 393. [CrossRef] [PubMed]
21. OpenCV. OpenCV: Optical FLow. 2021. Available online: https://docs.opencv.org/3.4/d4/dee/tutorial_optical_flow.html (accessed on 17 February 2021).
22. Nvidia. Jetson Nano. 2021. Available online: https://developer.nvidia.com/embedded/jetson-nano (accessed on 17 February 2021).
23. Insta360. Insta360 Air. 2021. Available online: https://www.insta360.com/es/product/insta360-air (accessed on 17 February 2021).
24. Bochkovskiy, A.; Wang, C.Y.; Liao, H.Y.M. YOLOv4: Optimal Speed and Accuracy of Object Detection. *arXiv* **2020**, arXiv:2004.10934.
25. Jocher, G.; Stoken, A.; Borovec, J.; NanoCode012; ChristopherSTAN; Changyu, L.; Laughing; tkianai; yxNONG; Hogan, A.; et al. ultralytics/yolov5: v4.0 - nn.SiLU() activations, Weights & Biases logging, PyTorch Hub integration. *Zenodo* **2021**. [CrossRef]
26. Farneb, G. Two-Frame Motion Estimation Based on. *Lect. Notes Comput. Sci.* **2003**, *2749*, 363–370.
27. MongoDB. The Most Popular Database for Modern Apps. 2021. Available online: https://www.mongodb.com (accessed on 15 February 2021).
28. Khurana, U.; Turaga, D.; Samulowitz, H.; Parthasrathy, S. Cognito: Automated Feature Engineering for Supervised Learning. In Proceedings of the IEEE International Conference on Data Mining Workshops (ICDMW), Barcelona, Spain, 12–15 December 2016; pp. 1304–1307. [CrossRef]
29. Wang, S.C. Artificial neural network (ANNs). In *Interdisciplinary Computing in Java Programming*; The Springer International Series in Engineering and Computer Science; Springer: Boston, MA, USA, 2003; Volume 743. [CrossRef]
30. Browne, M.W. Cross-Validation Methods. *J. Math. Psychol.* **2000**, *44*, 108–132. [CrossRef] [PubMed]

MDPI
St. Alban-Anlage 66
4052 Basel
Switzerland
Tel. +41 61 683 77 34
Fax +41 61 302 89 18
www.mdpi.com

Applied Sciences Editorial Office
E-mail: applsci@mdpi.com
www.mdpi.com/journal/applsci